STP 1107

Water in Exterior Building Walls: Problems and Solutions

Thomas A. Schwartz, editor

ASTM Publication Code Number (PCN)
04-011070-10

ASTM
1916 Race Street
Philadelphia, PA 19103

Library of Congress Cataloging-in-Publication Data

Water in exterior building walls : problems and solutions / Thomas A. Schwartz, editor.
    p. cm. — (STP ; 1107)
   Papers presented at a symposium, held in Dearborn, Mich., Oct. 25-26, 1990.
   "ASTM publication code number (PCN) 04-011070-10."
   Includes bibliographical references and indexes.
   ISBN 0-8031-1409-5
    1. Dampness in buildings—Congresses. 2. Exterior walls—Congresses. I. Schwartz, Thomas A. (Thomas Alan), 1951-II. American Society for Testing and Materials. III. Series: ASTM special technical publication ; 1107.
TH9031.W37   1991
603'.892—dc20   91-4483
                                                                                                                         CIP

Copyright ©1991 AMERICAN SOCIETY FOR TESTING AND MATERIALS, Philadelphia, PA. All rights reserved. This material may not be reproduced or copied, in whole or in part, in any printed, mechanical, electronic, film, or other distribution and storage media, without the written consent of the publisher.

Photocopy Rights

Authorization to photocopy items for internal or personal use, or the internal or personal use of specific clients, is granted by the AMERICAN SOCIETY FOR TESTING AND MATERIALS for users registered with the Copyright Clearance Center (CCC) Transactional Reporting Service, provided that the base fee of $2.50 per copy, plus $0.50 per page is paid directly to CCC, 27 Congress St., Salem, MA 01970; (508) 744-3350. For those organizations that have been granted a photocopy license by CCC, a separate system of payment has been arranged. The fee code for users of the Transactional Reporting Service is 0-8031-1409-5/91 $2.50 + .50.

Peer Review Policy

Each paper published in this volume was evaluated by three peer reviewers. The authors addressed all of the reviewers' comments to the satisfaction of both the technical editor(s) and the ASTM Committee on Publications.

The quality of the papers in this publication reflects not only the obvious efforts of the authors and the technical editor(s), but also the work of these peer reviewers. The ASTM Committee on Publications acknowledges with appreciation their dedication and contribution of time and effort on behalf of ASTM.

Printed in Baltimore, MD
February 1992

Foreword

This publication, *Water in Exterior Wall Systems: Problems and Solutions*, contains papers presented at the symposium of the same name, held in Dearborn, MI on 25–26 Oct. 1990. The symposium was sponsored by ASTM Committee E-6 on Performance of Building Constructions. Thomas A. Schwartz of Simpson, Gumpertz & Heger in Arlington, MA presided as symposium co-chairman and is the editor of the resulting publication.

Contents

Overview—THOMAS A. SCHWARTZ — vii

Water Intrusion in Barrier and Cavity/Rain Screen Walls—M. F. WILLIAMS AND B. L. WILLIAMS — 1

Design and Construction of Watertight Exterior Building Walls—S. S. RUGGIERO AND J. C. MYERS — 11

Diagnosing Window and Curtain Wall Leaks—R. J. KUDDER AND K. M. LIES — 40

Water Vapor Behavior in Exterior Insulation and Finish Systems—R. G. THOMAS, JR. — 53

The Practical Use and Potential Limitations of Exterior Insulation and Finish System Materials as an Exterior Building Envelope—W. R. FRENCH — 64

In-Situ Testing of the Structural Integrity of Exterior Insulation and Finish Systems—W. R. FRENCH — 79

Heat and Moisture Transport Through a Glass-Fiber Slab with One Side Subject to a Freezing Temperature—Y.-X. TAO, R. W. BESANT, AND K. S. REZKALLAH — 92

Comparison of a Wall Moisture Model with Field Data—H. S. JHINGER, G. K. YUILL, AND T. HAMLIN — 105

Vapor Control and Psychometric Monitoring in Exterior Walls—R. J. KUDDER AND K. R. HOIGARD — 124

Methods for Identifying Sources of Moisture in Walls—H. R. TRECHSEL — 138

Establishing Appropriate Field Test Pressures for Investigation of Leakage Through the Building Envelope—G. G. COLE AND T. A. SCHWARTZ — 150

Tracing Roof and Wall Leaks Using Alternating Electric Fields and Vapor Detection—J. C. MAY AND J. M. VASSILIADES — 160

Sealant Joint Design—C. BEALL — 165

Water Infiltration in Hawaii and Ensuing Construction Litigation—A. C. YANOVIAK 182

Prevention of Water Penetration Through Exterior Door Systems—D. W. KEHRLI 201

Computer Simulation of Wall Condensation Problems—A. R. CARLSON 210

Author Index 229

Subject Index 231

Overview

Nothing in our known universe is at once as constructive and destructive as water. It supports life while it turns stone to sand. It provides energy and delivers essential services to our built environment, while simultaneously degrading the materials in that built environment by corrosion, erosion, dissolution, decay, and expansion due to freezing. The papers presented in this Special Technical Publication (STP) explore both the problems associated with water in the exterior walls of buildings and the solutions to those problems.

The architectural and engineering journals are replete with stories about the adverse consquences of uncontrolled water in building walls; major overhauls of recent construction due to persistent water penetration, corrosion of light gage metal components, decay of organic materials, and damage to interior finishes. Our courtrooms are deluged with dissatisfied owners seeking restitution for building envelope defects. The source of these performance deficiencies include the following:

- wall designs that fail to recognize and accommodate the movement of moisture in both its liquid and vapor forms;
- installation workmanship that fails to implement the wall design faithfully; and
- materials and material systems that fail to deliver their advertized performance.

Correcting structures, once built, generally requires creative and usually expensive remedial measures. Anticipating and correcting problems, before construction, is much easier and certainly less costly. Problem prevention requires unwavering respect for the laws of physics, meticulous attention to detail, and a healthy appreciation for the lessons of history.

Despite constant changes in materials and methods, the construction of watertight, durable building enclosures relies now, as it has for decades, on the prudent application of building technology. As in politics, we must be students of history or we are doomed to repeat it. The consequence of ignoring the lessons of our past experience is highlighted by an article in *Engineering News Record* entitled "How to Build Leaky Brick Walls with Good Materials," which suggests the following elements to help assure water problems in masonry walls:

- Combine materials of varying coefficient of expansion to assure that cracks will develop through the veneer.
- Keep the edge of flashings 1/2 in. or more back from the face of the wall so that the outside appearance of the wall is not marred.
- Lap or loose lock, but do not solder, the laps of metal flashings.
- Do not turn up inboard edge of flashing and do not seal the flashing around penetrations and termination points.

- In laying brick, butter the outer edge with mortar, so that the finished wall looks good, but do not provide full head and bed joints.

This article was published in 1938. Those involved in the investigation of leaky brick walls in the 1980's have seen striking similarities in the current practice with that of the 1930's. Without action to improve the state of our art, the 1990's are unlikely to provide a significantly different experience.

Solutions to the problems of water in exterior building walls are at hand, but problem recognition, the first step in prevention, has proven evasive. The papers in this STP go a long way in both recognizing the problems and providing methods to solve them.

The papers included herein present diverse, and sometimes controversial opinions on the subject. A contrast of opinions is expected, given the disparate professional experiences of the authors. I welcome this diversity as an essential element in the development of consensus and improved practice.

Thomas A. Schwartz, Editor
Simpson Gumpertz & Heger, Inc.
Arlington, MA

Mark F. Williams and Barbara Lamp Williams

WATER INTRUSION IN BARRIER AND CAVITY/RAIN SCREEN WALLS

REFERENCE: Williams, M. F., and Williams, B. L.,"Water Intrusion in Barrier and Cavity/Rain Screen Walls," Water in Exterior Building Walls: Problems and Solutions, ASTM STP 1107, Thomas A. Schwartz, Eds., American Society for Testing and Materials, Philadelphia, 1991.

ABSTRACT: Exterior walls are designed and constructed using barrier or cavity/rain screen wall principles. Exterior Insulation and Finish Systems (EIFS) are typically constructed as barrier walls; masonry is often constructed as a cavity wall. These wall systems are discussed along with common deficiencies that allow water intrusion to occur.

KEYWORDS: barrier wall, brick veneer, cavity/rain screen wall, Exterior Insulation and Finish Systems (EIFS), unit masonry.

INTRODUCTION

Exterior building walls are designed and constructed to modify or exclude environmental forces from building interiors. These forces include heat/light, wind, sound, and moisture in the form of water vapor, rain, and ice. This paper discusses two exterior wall concepts—barrier and cavity/rain screen—as they relate to water. Each concept requires special attention to design and construction methods in order to achieve a properly functioning exterior wall. A prerequisite for successful exterior walls is sound architectural and engineering practices to accommodate reasonably expected loads and movement. This paper assumes that these typical considerations have been fully addressed; the specific points discussed are additional considerations.

Mark F. Williams and Barbara Lamp Williams are president and vice-president, respectively, of Kenney/Williams/Williams, Inc., a building diagnostics firm located at 945 Tennis Avenue, Maple Glen, PA 19002.

A barrier wall is designed and constructed to shed all water, thereby preventing any moisture from penetrating beyond the outermost surface and into the wall itself. Barrier walls are essentially moisture-tight constructions. Beneath the barrier's outer skin, the underlying materials and building interior are kept dry.

A barrier wall does not provide for the management of water that inadvertently passes beyond its outer skin. Once water passes the outer surface, degradation of the wall and its function follow. Single-wythe masonry, architectural precast concrete, and Exterior Insulation and Finish Systems (EIFS) are typically designed as barrier walls.

Cavity/rain screen walls also shed water at the outer face; however, this wall concept anticipates some degree of water intrusion by including internal drainage provisions. Specifically, an air space situated between the inner and outer wall components forms a "cavity." The dimension of this air space depends on the version of the cavity/rain screen concept implemented in the wall.

In the cavity wall version, the air space, detailed with flashing and drainage features, returns water to the building exterior. In its most basic form, the air space relies on the force of gravity to collect and redirect water to the building exterior. To function properly, the space must be unobstructed. Any blockage or obstacle within the space hinders the collection and proper discharge of water.

In the rain screen version, the air space serves as a means of wall drainage as well as a means of equalizing air pressure in the space. An air-tight barrier is placed within the inner layer of the wall construction. The air barrier and specialized vents equalize pressure differences between air inside the cavity and outside the building.

Since the cavity version of this concept does not attempt to equalize air pressure, a wider air space is generally detailed. This wider space eases construction and helps reduce cavity obstructions. Specifically, the wider space accommodates materials such as unit masonry and mortar that may vary in their dimensions or placement in the wall. Alternatively, the rain screen version necessitates a narrower, compartmentalized air space. This reduces the volume of air in the space and increases the rate of pressure equalization.

EIFS BARRIER WALLS

Most EIFS claddings are barrier walls: sealed systems that are intended to shed exterior water, stop water intrusion, and keep the building interior dry. EIFS are highly integrated composite systems. They incorporate several material layers: finish coat, base coat, reinforcement, and insulation board. EIFS coatings and reinforcement are applied to rigid insulation board that has been attached to the building substrate. To achieve proper system functioning, system materials are assembled to create a single, unified barrier in which all components act as one.

The primary EIFS barrier against water intrusion is the lamina, consisting of finish and base coats together with system reinforcement. The lamina acts as a waterproofing layer between the outside of the building and the insulation board. The base coat is the component most responsible for moisture protection. Its acrylic polymer constituents coalesce upon drying and form a continuous matrix capable of resisting water intrusion. The base coat also imparts flexibility to the system and resists cracking, with the aid of embedded fiberglass mesh. When properly installed, the mesh forms a continuous layer of system reinforcement that enhances the tensile strength, durability, and impact resistance of the lamina. Defects that originate in the lamina or develop in the lamina as a result of other system deficiencies can threaten the water-tight conditions essential to a properly functioning EIFS barrier wall. This is of particular concern when EIFS barrier claddings are adhesively attached to paper-faced gypsum substrates.

EIFS Barrier Wall Deficiencies

EIFS components are installed in successive steps to create a barrier wall with a durable weather-resistant lamina. The EIFS lamina is intended to function as a composite layer of integrated materials. Numerous application errors and omissions that cause cracking of the lamina, a common EIFS deficiency, can occur on a project.

Base coat thickness and the position of the embedded reinforcing mesh are critical factors which affect two central dimensions of system performance: moisture protection and impact resistance. When applied to the specified thickness, EIFS base coats can be reasonably effective moisture barriers. Non-cementitious base coats are typically more resistant to moisture transmission than those which incorporate portland cement. Whatever type of base coat is used, application to the specified thickness is essential for system waterproofing.

Thinly applied base coats do not provide a sufficient waterproofing layer and are more vulnerable to damage from external impact. It is difficult if not impossible to fully embed mesh in a thin layer of base coat. Fig. 1 shows a lamina cross-section in which the mesh is improperly positioned in a thin base coat. Fig. 2 shows a fully embedded mesh properly positioned in a base coat of specified thickness. Inadequate mesh embedment produces a poorly reinforced system with lower protection against impact. Thin base coats are subject to cracking and subsequent failure. Thick base coats also tend to crack, especially those which contain a high proportion of portland cement. Whether the base coat is too thin or too thick, once cracks develop in the lamina, they serve as routes for water intrusion.

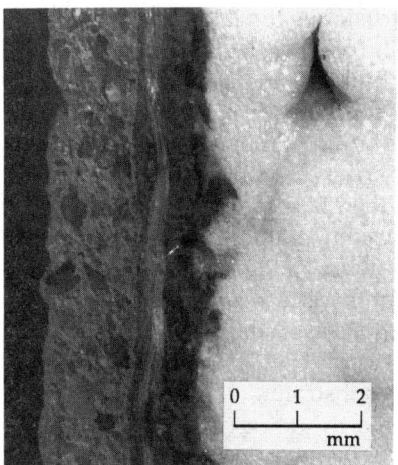

FIG. 1 — EIFS lamina cross-section showing improper mesh positioning.

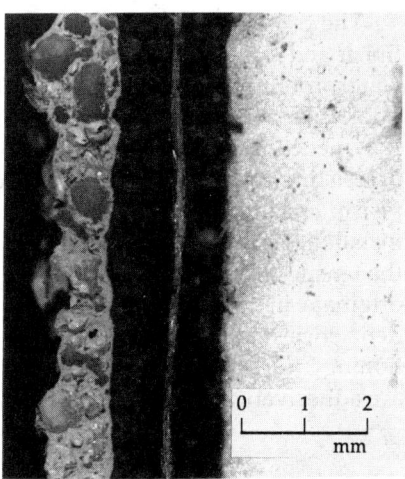

FIG. 2 — EIFS lamina cross-section showing proper mesh positioning.

Incorrect proportions of base coat components are a common reason for lamina cracking. Adherence to the manufacturer's recommended ratio of factory-blended base coat material and portland cement is critical to lamina integrity. If the amount of cement in the base mix is increased, or the polymer decreased, the resulting material will be more rigid and prone to cracking and spalling. The manufacturer's base coat formulation must give proper consideration to the granulometric curve of sand and filler materials used in lamina coatings. Well-formulated coatings include a full range of particle sizes and shapes to create compacted materials without voids.

The insulation board serves as a "substrate" for the components of the lamina. An essentially "jointless" surface should be achieved during insulation board installation to provide a satisfactory substrate for system coatings and reinforcement. Numerous deficiencies can originate in the insulation layer which cause the lamina to crack and jeopardize the EIFS barrier.

One of the most common deficiencies is poor abutment of insulation boards. Poorly abutted boards create gaps or breaks in the insulation surface—a surface which serves to support the overlying lamina. When boards are poorly abutted, there is no support for the lamina, resulting in lamina cracking along gapped board lines. Moisture inside the system can pass through board gaps and cause further deterioration of system components. In addition, the outline of board gaps can be detected through the finish coat at the system's surface, a condition which detracts from building appearance.

Poor board abutment can be caused by a number of factors, including poor quality insulation board. Depending on production methods and quality control procedures, insulation boards may vary in their dimensional consistencies: length, width, thickness, squareness and planar flatness. Varying quality boards are difficult if not impossible to abut tightly. Boards that are out of square produce gaps which fill with additional base coat and may result in cracks. Boards that vary in thickness produce step-in board faces and different lamina thicknesses.

Many improper application practices cause poor board abutment. For example, improper application of adhesive is a common factor in poor abutment. If adhesive is allowed to overrun the perimeter of insulation boards, tight board abutment is obstructed. Board edges should be wiped clean to remove excess adhesive prior to installing adjacent boards. If gaps occur during application, they must be fixed. It is mandatory that small slivers of insulation be inserted to fill gaps. Full rasping of installed boards, a recommended application practice, assists in leveling variations in thickness.

Window and door corners are areas of concentrated stress where cracking typically occurs. Careful installation practices are required. Saddle- or L-shaped boards should be installed at window heads to reduce the number of board joints at the corners of system openings. An incorrect practice is to install boards against door and window openings such that board joints align with window heads, sills, and jambs. If board joints coincide with corners, horizontal and vertical cracking can occur at these locations. Additional reinforcement in the form of diagonal mesh is required at the corners of openings.

Poorly designed sealant joints are a common cause of EIFS failure. Improper joint dimensions frequently contribute to joint failure. Joint depth on EIFS should not exceed joint width, with a maximum depth of 1/2 inch stipulated. Narrow joints or sealant material with insufficient movement capacity cannot successfully accommodate joint movement.

Proper sealant joint installation requires informed and skilled workmanship. Field applicators must verify that sealant substrates are clean, dry, and sound. Only specified materials should be used. Back-up rods, for example, should be of a closed-cell construction; opened-cell rods retain moisture which damages the system. Improperly mixed sealant or poorly tooled joints facilitate sealant failure and subsequent moisture intrusion. Water-tightness is the prime goal of sealant installation, and any condition or practice counter to this goal should be avoided.

Cohesive and adhesive sealant failures occur in EIFS claddings for the same reasons as in other types of exterior walls. These reasons include poor surface preparation, improper material specification, incorrect application, and poor tooling. Particular to EIFS is cohesive lamina failure at sealant joints. Certain finish coats tend to soften with prolonged exposure to moisture. Frequently, softening occurs where moisture has already penetrated the system barrier. For instance, moisture tends to collect on the upper portion of horizontal joints at backwrapped terminations which causes the overlying finish coat (the sealant substrate) to soften. The widespread EIFS practice of sealing to the finish coat contributes to this type of failure. Sealing to the base coat or to accessories are alternate approaches to sealant installation which avoid joint failure due to finish coat softening.

BRICK VENEER CAVITY WALLS

Brick veneer cavity walls are designed and constructed to shed most water at the outside wall surface. However, cavity walls anticipate that some amount of water intrusion will occur. Water that passes the outer wall is intercepted and redirected to keep the building interior dry. This is accomplished by means of an air space cavity which separates the exterior wythe of masonry from back-up wall construction.

Typical brick veneer cavity wall construction includes an exterior wythe of masonry veneer, a 2 to 4 inch wide cavity (or if rigid insulation is used, a minimum net cavity width of 2 inches should be provided), an inner wall substrate with weather barrier, insulation, and a vapor retarder, depending upon project specific conditions. The combined use of flashings and weep tubes above wall openings, at floor lines, and close to grade assists in redirecting water to the building exterior.

A sufficient number of metal ties, properly placed, is necessary to join the exterior masonry wythe to the substrate and transfer lateral loads between wall members. If the ties between masonry veneer and substrate are inadequate for any reason, flexural cracks in the wall surface will most likely occur. When compared to barrier wall construction, cavity wall construction is better equipped to handle water intrusion; however, it is less able to resist lateral loads because of its lower flexural strength.

Brick Veneer Cavity Wall Deficiencies

Brick veneer cavity walls should be designed and constructed to implement specific wall principles. Since the wall must shed water at the outside wall surface, appropriate grade masonry units are combined with a full bed of mortar. Poor mortar selection or mix can result in wall deficiencies. Poor application practices include improper mortar installation at head joints (vertical joints between masonry units) or poor tooling which allows excess water to penetrate the masonry face. Typically, the most critical interface is the juncture between mortar and masonry units. A reasonably water-resistant joint is obtained by covering the entire joint between masonry units with mortar that is well-compacted and tooled in a concave profile. Mortar that is stronger in compression than needed to meet the project's structural requirements should not be used.

Cavity wall construction anticipates that some water will penetrate the outer wall surface. A cavity air space is provided to help collect this unwanted water. Water infiltration is facilitated by cracks within the wall or a poor bond between mortar and masonry units. Water which enters the cavity system is drawn into the wall because of air pressure differences between the cavity and the exterior.

Although the cavity concept appears straightforward, many designs and construction assemblies simply do not implement the details necessary to accomplish the overall principle. Given the important role of the cavity, sufficient width and clear passage are critical for proper functioning. A narrow cavity is difficult if not impossible for a mason to keep clean during construction (Fig. 3).

Mortar which is not removed from the inside surface of a brick veneer wall will bridge narrow cavities. Given the weight and material properties of fresh mortar, these bridges tend to drop from the wall surface and fill the bottom of the cavity space. Mortar droppings often block access to weep tubes thereby precluding the discharge of water from the cavity (Fig. 4). In a worst case condition, we have observed the cavity space behind soldier courses filled intentionally with mortar, presumably to help stabilize the vertical masonry units during construction. As a result, the cavity and weep tubes were completely blocked and rendered ineffective. Mortar droppings should be removed to prevent the movement of water by capillary action across the cavity. At the cavity base, the placement of small, river-washed stones will generally prevent the minor accumulation of mortar droppings from blocking the interior ends of weep tubes.

Aside from water collection, redirection of water to the building exterior is accomplished with flashing. Flashing membranes should be flexible and durable. Materials which become rigid and brittle during cold weather should be avoided. Flashings should continue vertically 6 to 9 inches above the horizontal plane, and extend horizontally to a point outside the exterior wall surface. Without proper extension beyond the outside face of the wall surface, water that collects in the cavity may run underneath the flashing and return to the wall assembly (Fig. 5). Relatedly, sealant should never be placed over the front edge of flashing. If relieving angles are used, and sloped toward the building interior, water can easily run under flashing which has not been properly extended. Sealing between pieces of flashing and providing adequate end dams at flashing terminations are installation practices which assist in proper water redirection.

The installation of adjacent systems by various trades must be coordinated with masonry construction. For example, powder actuated fasteners used to attach the head receptor of a window system may inadvertently pierce the flashing membrane (Fig. 6). This installation defect depends upon the length of the fastener as well as the strength of the powder actuated fastener load.

The integration of the window system with masonry construction tolerances is important. For example, brick veneer construction typically includes some variation in the horizontal and vertical planes. By contrast, window extrusions are typically straight, with negligible variation. If the design intent is to have a window system flush with the face of the masonry wall, it is very likely that the window

FIG. 3 — Narrow cavity with mortar accumulation.

FIG. 4 — Weep tube plugged with mortar in cavity wall.

FIG. 5 — Insufficient extension of flashing and accumulated mortar in cavity space.

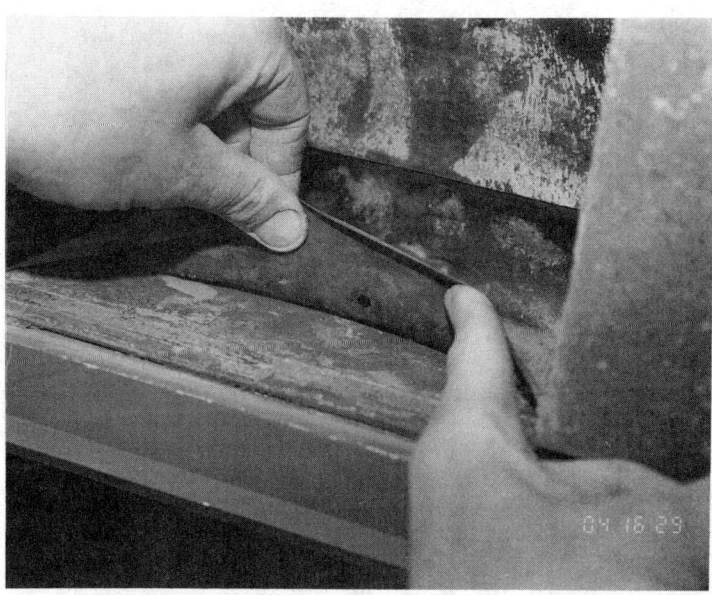

FIG. 6 — Penetration of flashing by powder actuated fastener in cavity wall.

system will end up outside the wall plane due to the wide construction variances associated with unit masonry. The planar relationship between systems then relies upon sealant material to exclude all moisture and provide a transition between window receptor and unit masonry. A reasonable solution to prevent this type of problem places the window extrusion approximately 3/4 inch behind the wall face, thereby accommodating construction variances.

CONCLUSION

Barrier and cavity/rain screen walls represent two different approaches to dealing with the common threat of water intrusion. EIFS barrier walls rely on a composite assembly of polymer materials to impart water-resistant properties and flexibility to the exterior wall. Brick veneer cavity walls utilize traditional materials which are assembled in two layers and separated by an air space with appropriate through-wall flashings at penetrations and terminations. EIFS barrier walls and brick veneer cavity walls, if properly designed and constructed, can realize satisfactory performance with respect to shedding or managing water intrusion.

Stephen S. Ruggiero and James C. Myers

DESIGN AND CONSTRUCTION OF WATERTIGHT EXTERIOR BUILDING WALLS

REFERENCE: Ruggiero, S. S., and Myers, J. C., "Design and Construction of Watertight Exterior Building Walls," Water in Exterior Building Walls: Problems and Solutions, ASTM STP 1107, Thomas A. Schwartz, Ed., American Society for Testing and Materials, Philadelphia, 1991.

ABSTRACT: In this paper, we review basic principles of design for two fundamental categories of wall waterproofing systems, barrier walls and cavity walls. We then examine the application of these principles to construction of various types of wall systems, including single-wythe masonry veneers, precast concrete wall panels, glass/metal curtain walls, and exterior insulation and finish systems. For each system, we discuss key features for reliable and durable waterproofing, common problem areas that we have encountered, and remedial options for leaking systems.

KEYWORDS: water leakage, barrier walls, cavity walls, flashings, masonry veneers, precast concrete wall panels, curtain walls, exterior insulation and finish systems, wall repair, sealant joints

Exterior building walls generally consist of an exterior veneer or cladding that provides the weathering surface of the building, a back-up that provides structural support for the veneer, and an interior finish applied to the back-up. Buildings from the early 1900's have relatively massive exterior walls with multiple layers of thick absorptive materials separating the exterior surface from the interior finishes. The articulation of the exterior facade promoted drainage away from wall openings; designs incorporated secondary waterproofing barriers or built-in flashings for long-term performance.

Current trends in exterior wall design have led to increasingly thin, lightweight veneers with little separation between exterior surfaces and interior finishes. In many cases, secondary barriers and through-wall flashings are absent from the design and surface water flows freely over exposed joints and wall openings. As a result, the occurrence of exterior wall leakage problems has increased, including consequential degradation from such leakage, such as deterioration or corrosion of hidden wall components and damage to interior finishes, within the first few years of service. Water vapor may condense within exterior walls and deteriorate components, but this paper does not address this source of moisture.

Mr. Ruggiero is an Associate and Mr. Myers is a senior staff engineer at Simpson Gumpertz & Heger Inc., Consulting Engineers, 297 Broadway, Arlington, MA 02174

In this paper, we examine two fundamental approaches to waterproofing exterior walls:

- "barrier wall" construction – use the exterior surfacing as the sole waterproofing barrier, or

- "cavity wall" construction – provide a waterproof barrier behind the exterior surfacing to collect and drain water that penetrates the veneer back to the exterior.

This paper reviews basic principals of design for waterproof wall construction and then examines applications of these principals to construction of various types of wall systems. It discusses key features for reliable and durable waterproofing in each system, as well as common problem areas that we have found in our field investigations and remedial solutions for leaking walls.

RAINWATER ON EXTERIOR WALLS

A sound approach to aid in the waterproofing of exterior walls is to shield them from rain, such as by using cornices, overhangs, belt courses or similar features. Unfortunately, the effectiveness of this approach is limited to low-rise construction. As building height increases, rain accompanied by even slight wind tends to wet the wall surfaces despite such shielding features. The following examines two categories of exposure to the elements: rainwater flow over the wall surface with and without the driving pressure of wind.

Gravity-Induced Water Flow

Water that flows down exterior walls soaks into absorbent surfacing materials and flows into cracks or other openings in or between the various wall components. Gravity, surface tension, and capillary action allow water to penetrate the openings even when wind and its driving pressure are absent. Our experience in evaluating and testing various wall systems is that much of the leakage can be replicated by allowing water to flow over the wall system without application of a differential pressure across the wall, i.e., wind pressure. While wind is a key element in the waterproofing design of wall systems, the designer should reduce the exposure of the wall components and joinery from water flow due to gravity.

Providing slight outward slopes to horizontal surfaces avoids ponding of water and directs water away from the wall and joinery. Shingling or overlapping materials at joints in the direction of water flow reduces the severity of joint exposure to water.

Critical areas of the wall should be shielded from the flow of water down the wall. Setting windows back from the face of the wall is an example of this approach. Providing an exposed drip edge on metal extrusions or flashings above horizontal joints reduces the exposure of the joints to rain water. Projecting subsills that extend beyond window jambs and continuous ledges or belt courses shed water away from vulnerable joints, and break up concentrated water flows and spread them more evenly over the wall surface.

The above approaches use physical building features to provide permanent protection of the vulnerable areas with little maintenance requirements. However, such features do not insure watertight wall construction; the design must account for the effects of wind-driven rain.

Wind-Driven Rain

Wind creates two types of driving forces on rainwater: the momentum of the raindrops (splashing) and differential pressure across the exterior wall. The momentum of wind-driven raindrops allows them to penetrate openings approximately 8 mm (0.25 in.) or wider. Narrower openings cause the raindrop to shatter with less penetration into the opening. Wind pressures can counteract the effects of gravity and cause water to "flow uphill", but most importantly, wind creates a pressure differential across the wall that forces water through cracks and openings in the wall cladding. Water held within the cladding due to capillary forces will flow readily toward the interior under small differential pressures. Designers need to consider these forces that act on wall systems, particularly on taller buildings and those in windy locations, such as shorelines of lakes and oceans. Reference # 1 contains details on calculations of wind pressures on buildings.

FUNDAMENTAL DESIGN CONCEPTS FOR WATERPROOFING EXTERIOR WALLS

Conceptually, wall waterproofing systems fall into two categories, depending upon the means by which they control rainwater and the driving forces discussed above. Barrier walls rely on the exterior cladding and surface seals at joints to prevent water penetration to the interior. Cavity walls rely, in part, on the cladding to shed rainwater, but include a back-up waterproofing system to collect water that penetrates the cladding surface and drain it back outside. These two categories provide a convenient basis for examining the waterproofing fundamentals discussed below; however, many wall systems consist of combinations or variations of these two types. In a subsequent section of this paper, we discuss the application of these fundamentals to commonly used wall systems.

Barrier Walls

Barrier wall designs require that the exterior wall materials and joinery block passage of all water at the exterior face of the wall. These systems typically have no waterproofing redundancy and little tolerance for construction variations and defects. Key factors in the performance of barrier walls are discussed below.

The basic cladding element must be relatively impermeable and cannot develop through-cracks in the course of weathering and reacting to thermal or moisture cycles. Some materials contain more redundancy than others and reduce the chances of leakage. For example, multi-wythe brick masonry can tolerate some deficiencies in construction of one of the wythes without creating through-cracks. Some materials absorb and contain limited moisture without significant material deterioration or leakage to the interior.

Joints in the wall system at openings or between cladding elements must be sealed with materials that do not split or debond from the cladding. The cladding must be continuous and uncracked along the sealant bond line. Typically, cladding joints are sealed during construction (in the field) with liquid-applied sealants. It is unreasonable to expect the application of these materials to be perfect. Substrate surfaces must be sound and uncracked and then cleaned and prepared for sealing. Joint back-up materials need to be positioned properly, and sealant materials must be mixed in some cases and then applied. The sealant materials must withstand joint movement and weathering without deterioration. Given all of these factors and variables, some deficiencies in the joint seals are likely to occur both upon initial installation and as the sealant ages. Under the best circumstances, the number of deficiencies are small and leakage is not widespread, but maintenance of the seals is necessary to avoid increased leakage. In field surveys and testing, we have found significant leakage problems in buildings with single joint seals that contain defects along as little as 1% of their length. This does not provide much allowance for variability in construction of such joints. A common method to improve the watertightness of sealant joints is to provide two seals in one joint. A later section discusses this approach.

Incorporating shielding elements within the cladding to protect the joints can improve barrier wall performance significantly. Overlapping the wall elements at joints, recessing the seals and windows from the face of the wall, and providing overhangs and drip edges are examples of such features. Recent trends in wall design eliminate such features and set glazing and joint seals flush with the exterior surface with little shielding from rainwater and from the deteriorating effects of UV radiation on organic sealants.

Elements within wall openings, such as windows, must be watertight and cannot leak from frame corners or face joints. Windows typically contain joints between the horizontal and vertical framing members that are sealed with gaskets or liquid-applied sealants. For reasons discussed above, corner seals that are constructed with liquid-applied sealants are not likely to be watertight. In addition, handling and installation of the window frame can disturb or break these seals (Figure 1). For these and other reasons, it is prudent to install a flashing, such as a sheet metal pan, along the bottom of the window to collect leakage through the window glazing or frame joints and direct it back to the outside (Figures 2 and 3). Reference # 2 provides more detailed information on window sill flashings. Many barrier walls do not incorporate such a flashing in keeping with the concept that the surface seal is the only defense needed against water penetration. In many cases, this results in water leakage into the building.

Wall openings interrupt the cladding. Some barrier walls, such as multi-wythe brick masonry, may absorb and contain some water within the cladding. As this water seeps down within these materials, it can leak to the inside at the top of the wall openings, unless a flashing is installed in this location to collect water and drain it to the exterior.

Our experience is that barrier walls generally are problematic because the combination of imperfect average workmanship and degradation of materials by weathering result in deficiencies in the barrier that allow some water leakage. The extent and nature of leakage problems that develop depend heavily on the types of materials used, the quality of workmanship, and the frequency of maintenance.

FIGURE 1. Corner of a horizontal sliding window frame. A thickness gage is inserted through the unsealed joint between the intersection of the sill track and jamb extrusions that provides a path for water penetration.

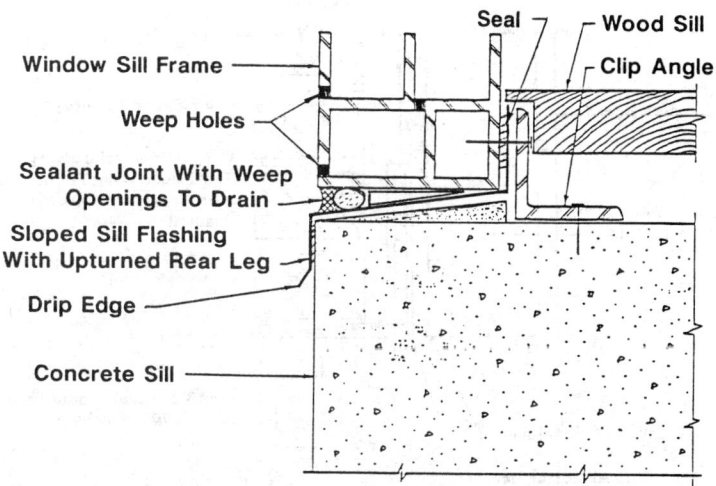

FIGURE 2. Schematic cross section of window sill flashing. The flashing collects water that penetrates the window, such as at corners (see Fig. 1), and drains it back to the exterior through weep holes. Window frame also has drainage ability.

FIGURE 3. Window sill flashing installation. Remedial lead-coated copper flashing has been placed in a masonry opening prior to setting the window. Note upturned inboard leg and end of flashing.

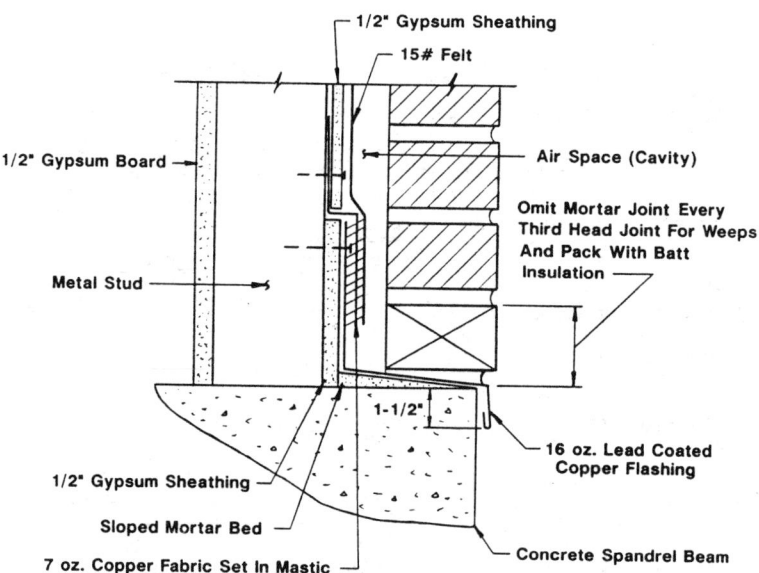

FIGURE 4. Through-wall flashing of brick veneer/steel stud wall.

Cavity Walls

The cavity wall concept differs fundamentally from the barrier wall concept, in that the exterior surfacing screens the rain from the waterproofing layer that is placed behind it, rather than acting as the sole barrier to water entry. This concept acknowledges, and accounts for, the inevitable penetration of some water through the exterior veneer and joinery. As such, it avoids some of the primary drawbacks of the barrier wall approach and can possess a high degree of reliability and durability. The details of cavity wall construction can take different forms depending upon the veneer type and back-up construction; its fundamental elements include the following.

The exterior veneer provides the initial barrier to water penetration. While the veneer is not expected to prohibit all water entry, it should not contain significant cracks, openings, or unsealed joints. Differential air pressure acts across this veneer and drives water through it.

An air-space isolates the inner, or back-up wall, from the exterior veneer. Water that penetrates the veneer flows downward in this cavity, minimizing any contact with the back-up wall construction. The width of the air-space varies depending upon the veneer materials and the likelihood of creating obstructions during construction of the veneer, but generally ranges from 2.5 to 5 cm (1 to 2 in.).

A continuous waterproofing layer should cover the back-up wall to shed any small amounts of water that inevitably cross the air space by splashing or by direct flow at cavity obstructions or at veneer anchor ties that span the cavity. Asphalt saturated felts, shingled with the flow of water, are commonly placed on the exterior face of the back-up wall. Because the veneer and cavity control much of the water and the veneer shields the cavity from wind-driven rain, the requirements for this waterproofing layer are much less severe than if it were exposed on the face of a building. The combination of a protective screen and a waterproofing layer provide significant redundancy in these systems with resultant long-term reliability.

Horizontal runs of through-wall flashings must be located at regular vertical intervals to collect the water that flows downward within the veneer and cavity space. The inboard end of the flashing should turn upward at the back-up wall and the wall waterproofing layer should shingle over it. The flashing should extend from the back-up wall, across the cavity, through the veneer, and terminate with an exposed drip edge at the front of the veneer to prevent water from running back underneath the flashing (Figures 4 and 5). Providing slight outward slope to the horizontal part of the flashing to promote drainage and avoid ponding on the flashing enhances reliability and durability. Sloped quick-set mortar beds or closely-spaced tapered shims beneath the flashing can provide such slope.

Along the length of the wall, the flashing needs to be continuous and seamed watertight at joints and corners. Expansion joints should be incorporated in continuous flashings that are made with rigid materials, such as sheet metal, to accommodate thermally-induced movement of the flashing and cladding. At terminations, the flashing should turn up and the corner should be sealed watertight to prevent water from draining off the end of the flashing and into the building. Weep openings are needed in the veneer at the flashing level to permit drainage of

FIGURE 5. Installation of remedial through-wall flashing on a steel relieving angle above strip windows in a brick veneer/steel stud wall. Note exposed drip edge formed at front of flashing.

FIGURE 6. Looking at the cavity side of a brick veneer from inside with the interior drywall and gypsum sheathing removed. Note the mortar obstructions in the cavity (A) and the water penetrating at debonded mortar joints during a watertest (B).

water from the flashing to the exterior. Size and spacing of these weep holes varies with the veneer materials.

Pressure-equalized design concept: An approach that is related to the cavity wall concept is pressure-equalized design, which provides an air barrier inboard of the veneer, instead of, or in addition to, a waterproofing layer. By preventing air penetration through the back-up wall and by sufficiently venting the cavity (air chamber) to the outside air, the pressure differential across the exterior veneer is reduced, or eliminated, during wind-driven rains, thus removing a primary driving force for water penetration. Essential elements for pressure-equalized systems include the following.

The air barrier must be continuous and properly sealed to all wall openings such as windows and doors. The air chamber is not simply a ventilated space. Because wind pressures vary considerably over the face of the wall, the air chamber should be compartmentalized to avoid air flow, and accompanying water flow, from high pressure to low pressure regions. The air barrier and its supporting wall, typically the back-up wall, must have adequate strength to resist wind loads on the building.

The exterior veneer serves as the primary rain screen or barrier to water penetration. However, the joints between veneer elements are left open to some degree to allow efficient pressurization of the air chamber behind the veneer. Wind-driven rain inevitably penetrates the open joint areas due to momentum of the raindrops. Back-up waterproofing layers are needed at the joints or the joints should be configured to control this form of penetration, e.g., ship-lap geometry. Internal drainage devices, such as through-wall flashings, are required at regular vertical intervals to collect water that penetrates the cladding and direct it back to the exterior.

APPLICATION OF WATERPROOFING DESIGN PRINCIPALS IN TYPICAL EXTERIOR WALL SYSTEMS

The following describes the composition of some common exterior wall systems, the application of the design principals discussed above, and key features to incorporate in the design and construction of these systems. We do not discuss all features or aspects of proper design and construction of these systems. We also present common problems with these systems, and remedial options for leaking systems.

Cavity Wall/Masonry Veneers

A typical masonry veneer wall consists of nominal 10 cm (4 inch) thick brick veneer with a 5 cm (2 inch) wide air space (cavity) that separates it from the back-up wall. Wire ties embedded in the veneer bridge the cavity and are attached to a back-up wall to stabilize the veneer against wind loads. A layer of felt waterproofing covers the back-up wall, i.e., concrete masonry units or gypsum sheathing board/steel stud wall. Mastics have been used to waterproof concrete masonry unit back-up walls, but these mastics can crack as the back-up moves in response to changes in thermal, moisture, and loading conditions.

Single-wythe masonry veneers must be designed as a cavity wall to properly control water penetration through the veneer. These walls contain many mortar joints and some inevitable brick-to-mortar separations due to normal material and construction variations that allow water penetration (Figure 6). In addition, some moisture may soak through these somewhat absorptive materials. Proper selection of masonry materials and complete filling of mortar joints can minimize, but not eliminate, water penetration through the veneer. The cavity wall approach is necessary to accommodate this inevitable water penetration.

An important aspect in the construction of these systems is maintaining a clear cavity and avoiding accumulations of mortar droppings on the through-wall flashings. Care in placing the mortar and setting the brick units can reduce the amount of mortar oozing out from the cavity-side of the mortar joint and falling into the cavity (Figure 6). See Reference # 3 for further details. A cavity width of 5 cm (2 inches) makes it easier to control the mortar and reduce droppings into the cavity than with narrower cavities. However, we have seen successful construction with narrower cavities.

Through-wall flashing is the most essential element to successful waterproofing of single-wythe veneers. Steel relieving angles that support the veneer are typically located at each floor level, and the flashings should be located on each angle to limit the accumulation of water within the wall cavity, as well as to limit the distance it travels, before being weeped to the exterior. In some cases, exposed concrete spandrel beams support the brick veneer. Through-wall flashing is necessary at these areas, particularly since the spandrel beam tends to funnel any water leakage directly to the interior floor.

Through-wall flashings are also required to protect the heads and sills of wall openings against water penetration, such as at windows. Flashings at the head of the window are absolutely essential for cavity wall construction to collect the water draining down the wall cavity above the window head. Many windows are placed directly below the veneer relieving angles, and the flashing that covers the angle serves to protect the window head as well (Figure 5). Avoid penetrating the flashing with fasteners used to anchor the window head.

If the windows are placed into separate, "punched" openings in the wall, a loose steel lintel typically supports the brick veneer above the opening. A head flashing should cover the angle and integrate with the back-up wall waterproofing. The flashing should extend beyond the sides of the opening and the ends of the flashing must turn up with watertight corners, i.e., bulk head or end-dam, to prevent water from flowing off the ends of the flashing and into the wall assembly. Aligning bulk heads with a head joint in the veneer allows the bulk head to extend outward to the face of the veneer.

Sill flashings are generally necessary to waterproofing window openings (Figure 3). When the interior face of the window frame aligns with the interior face of the veneer, then any leakage through the sill-to-jamb frame corner or around the window frame may flow into and be weeped out of the cavity reducing the importance of a sill flashing. We recommend use of sill flashings regardless of frame position, to protect against inadvertent transmission of water to the back-up at wood blocking or other "rough opening" materials and sill anchors. We have successfully created sill flashings that direct the water into the cavity, rather than extending the flashing through the veneer, to avoid the aesthetic impact of an exposed flashing and drip edge (Figure 7). Such an approach requires careful consideration of the path of

FIGURE 7. Flashings at a window corner in a brick veneer/steel stud wall prior to placing brick. Sill flashing (A) will drain into wall cavity behind brick veneer. Copper fabric jamb flashing (B) connects felt waterproofing to window jamb and laps over sill flashing.

FIGURE 8. PVC flashing installation on a steel relieving angle. Note failure on right side to extend PVC across the angle beneath the brick (A). Tip of bolt has cut a hole in the PVC (B). Left seam was lapped without mastic and unsealed (C). Note concealed front edge of flashing (D).

water flow within the cavity, the impact of additional water in the cavity, and the construction details. Avoid penetrating the horizontal portions of sill flashing with window anchors, and use fasteners with seals through the vertical legs of the flashing to reduce the severity of exposure of these penetrations to water (see Figure 2).

The sides or jambs of the opening also require some protection as the waterproofing layer terminates at this location and, therefore, provides an avenue for penetration of water that flows in the cavity. This area is particularly vulnerable when the brick veneer forms a ninety degree corner at the jambs, since the return edge of the brick may reduce the width of the cavity. We commonly find that mortar tends to accumulate in such areas and directs water against the window jambs. Many jamb flashing details are available, depending upon the type of window framing. We frequently use sheet metals or copper fabric flashing. These flashings need to integrate with head and sill flashings and shingle properly over the sill flashing (Figure 7).

Flashings must be constructed from durable materials that can withstand abuse during construction of the brick veneer. Sixteen ounce lead-coated copper and 26 ga stainless steel have superior strength, corrosion and staining resistance, and can be bent and soldered to form durable watertight geometries. The seams between sections of flashing can be soldered or strip flashed with uncured rubber sheet to provide continuity of the waterproofing. Also, these metals can protrude beyond the face of the veneer to form drip edges that protect vulnerable sealant joints at "soft joints" below relieving angles and at the heads of windows.

To facilitate the installation of through-wall flashing, particularly its integration with the gypsum sheathing or concrete block at the back-up wall, a two-piece flashing assembly consisting of copper fabric and lead-coated copper or stainless steel is convenient. The 7 oz copper fabric can be shingled into the gypsum sheathing or concrete block slightly above the flashing level and protrude from the sheathing. When the through-wall flashing is placed at a later time, the copper fabric is then lapped over the rear upturned leg of the flashing (Figure 4).

The design of the through-wall flashing should provide adjustment capability to move the flashing in or out to maintain a uniform exposure of the drip edge over the masonry. Turning up the rear leg less than ninety degrees allows such flexibility.

Common problems: Some flashing materials, such as lightweight copper fabric (less than 5 oz) and thin unreinforced polyvinyl chloride (PVC) roll flashing (less than 1 mm, 10 to 30 mils), are readily punctured and torn during construction of the brick veneer, including after they are mounted on the back-up wall when the wind slaps them against the building (Figure 8). These materials are not stiff enough to maintain formed shapes and are damaged by UV exposure. Therefore, they cannot be formed to provide an exposed drip edge. The final positioning and seaming is commonly done by the mason, not a waterproofing contractor. We frequently find that joints are lapped and unsealed, or not lapped at all, particularly at corners (Figure 8). Lack of coordination of the various trades and failure to integrate the wall components is a common problem with this system.

We have investigated a number of walls with leakage problems, in which PVC roll flashings have become brittle and developed cracks and splits. PVC is a rigid plastic, which is made flexible during manufacturing by the addition of oils and "plasticizers." Embrittlement due to plasticizer migration is common to all PVC materials and is a significant problem with the relatively thin PVC roll flashings. We have seen such PVC embrittle within two years of service, particularly where the flashing is under mechanical stress (e.g. where the flashing spans over an offset or where mortar has accumulated on the flashing).

Aggravating any flashing deficiencies is the common problem of mortar accumulation in the wall cavity (Figure 6) and on the flashing, and use of small, widely spaced weep holes. Keeping the cavity clear requires close attention by the mason. For weepholes, we suggest providing open head joints filled with glass fiber batt insulation to maximize drainage from the flashing and prevent insect entry.

A common weakness in some flashing designs is to terminate the flashing behind the face of the veneer, concealing it from view (Figure 8). This practice can allow the water to run back underneath the flashing as it tries to drain from the cavity. This water then can either be conducted inside directly, such as with exposed concrete spandrel beams, or it can collect on steel support angles. The water on the steel angles can corrode the angles, and it tends to run along the angle and leak into the back-up wall at joints in the angle or at the ends of the angle. Extending the flashing through the wall and providing an exposed drip edge avoids this problem. An alternative is to fully adhere the flashing to prevent water from running underneath it. This alternative is not as reliable as the drip edge approach, since it relies on the quality and durability of the adhesive installation and it requires a joint-free substrate for continuous adhesion.

We find some projects where head or sill flashings are not included and leakage results. Also, a number of leakage problems result from poor flashing design, e.g., missing end-pans, unsealed joints and corners, penetrations by fasteners that anchor the window head or sill, etc. Flexible flashings should be folded to form a watertight corner, and should not be cut at the corner.

Remedial options: Remedies for leaking masonry veneers fall into three general categories - surface coatings, flashing replacement, or replacement of the wall. We have found that attempts to eliminate leakage through use of surface sealers and water repellents, such as siloxanes, are not generally successful. They are tried frequently because they are low cost and low disruption options compared to flashing or wall replacement. This approach does not treat the root cause of the leakage problems, which is usually defects in the flashing. Instead, the sealer attempts to reduce the volume of water penetrating the veneer and reaching the flashing, in a sense, reverting to barrier wall construction. Sealers can reduce the surface absorption and capillary draw of masonry walls and, thereby, reduce the amount of water penetrating the veneer via these paths and reaching the flashings. Generally, however, the sealers do not seal the separations or cracks between the mortar and the masonry units and water will continue to penetrate the veneer via these paths. In many masonry veneers, these separations are the predominant source of water entry, and leakage will continue despite sealer application. Frequent reapplication is necessary over the life of the building to maintain the effectiveness, if any, of the sealer.

A common repair is to replace defective flashings (Figure 5). This requires removing several courses of masonry at the flashing level in "leg and leg" fashion. One to two meter (3 to 6 ft) sections of masonry are removed, while adjacent sections are left in place between these areas or shoring is installed to temporarily support the veneer above. The flashing is repaired or replaced in the areas of removed masonry. The masonry is replaced, and the process is repeated until all of the flashing is repaired or replaced. At the same time, the base of the cavity can be cleared of any mortar obstructions, and proper weep holes incorporated.

Generally, the decision to replace the entire wall is due to other deficiencies beyond leakage problems, such as inadequate veneer ties, defects in the masonry materials, or deterioration of the back-up wall components from the on-going leakage.

Precast Concrete Wall Panels

These wall systems typically contain large prefabricated wall panels that are attached to the structure at a few discrete points to resist gravity and wind loads. There are horizontal and vertical joints between the panels. Strip windows, i.e., a continuous horizontal band of windows, are common with this system. Typically, a steel stud wall behind the panel supports the interior finishes, or metal furring is attached to the interior face of the panel to receive interior finishes.

Precast concrete panel systems can be barrier or cavity walls, including pressure-equalized designs. Barrier wall construction results from sequencing the wall erection such that the panels enclose the structural frame quickly and in advance of interior wall construction. Consequently, access to the exterior face of the interior walls cannot be achieved for installation of a waterproofing layer or air barrier.

To properly implement cavity wall construction, the back-up wall must be installed before the panels, and, therefore, must be capable of resisting wind loads during construction. Installation of the waterproofing layer, particularly the seal around panel attachment anchors, and the continuous through-wall flashings with associated seams and transitions typically requires access from the exterior and coordination with panel erection so that these operations can be completed as each panel is erected. Prefabrication and mounting of the flashing before erection can help reduce coordination problems. All of these factors increase the cost of the project and can reduce overall floor space. Consequently, the majority of precast panel wall systems are designed as barrier walls.

Architectural precast concrete wall panels can develop full-depth cracks, commonly at the re-entrant corners in the panels. Cracking is more common in sandwich panels, i.e., those with insulation placed within the panel during casting, than in solid concrete panels, due to greater thermal gradients across the panel depth. Proper quality control in manufacturing and handling during erection can reduce full-depth cracks in the field of these panels. Using panels with simple geometries, i.e., rectangular without "punched" openings, and simple anchorage arrangements that avoid restraint of thermally-induced bowing further reduces the likelihood of cracking. Accordingly, solid precast concrete panels can provide a fairly effective, but not always perfect, barrier against water penetration. Unlike some other wall systems that rely on light gage steel framing and gypsum sheathing for attachment, precast concrete panel systems rely on relatively thick steel angles and similar substantial materials for structural support and the system can tolerate some water entry without rapid structural deterioration.

The joints between panels can be significant sources of water penetration. Several options for waterproofing the joints are available. The simplest form of protection at the joinery consists of a single line of sealant material, typically a liquid-applied sealant, placed in a butt joint at the face of the panels. This approach is not as reliable as other methods because some water inevitably penetrates these single sealant joints and the butt joint configuration allows direct transmission to the interior. Half-lapped or ship-lapped joints can improve single joint seal performance, particularly when the sealant is recessed within the joint (Figure 9). Further, some panel edge geometries can protect the recessed sealant by including preformed drip edges at horizontal joints (Figure 9) and raised shoulders along vertical joints to reduce sideways flow of wind-driven water over the sealant (Figure 10). We have found many installations where polyurethane or polysulfide sealants shielded from prolonged UV exposure are in much better condition, i.e., less surface crazing, splitting, debonding and hardening, than sealants placed on the face of the building under direct UV exposure.

Joint reliability can be improved further by installing two seals in each joint, one near the face of the panel and the other set some distance behind the outer joint. This two stage approach provides redundancy in the system and protects the inner seal from the elements. This approach requires the installation of weep openings in the exterior seal to allow water contained by the inner seal to exit the cavity between joint seals. At vertical joints, the inner seal must turn out to the plane of the exterior seal at regular intervals to force water out of the joint (see Figure 11). This termination requires care in detailing and construction. Some outward slope or offset joinery should be incorporated in the horizontal panel joints to promote drainage. Failure to provide these weep openings results in water trapped within the wall and ponding against both seals; this accelerates deterioration of the sealant material and its bond to the substrate.

A more reliable approach is to incorporate a horizontal flashing at the base of the vertical joints. This avoids the problem-prone weep hole detail in the two-stage approach and reduces reliance on the horizontal sealants. Flashings can be incorporated easily with strip window systems, because continuous window sill and head flashings can be installed after the panels are erected. We recommend using metal flashings, as shown in Figure 12, to drain water from the system.

Panel openings, such as at "punched" windows, require sill flashings. The panel edges can be configured to shield the perimeter joints, direct the flow of water away from the opening, and restrict the transmission of water to the interior, such as with steps or jogs in the panel edge inboard of the sealant joint. Head flashings generally cannot be installed, because the ends of the flashing cannot turn up into the solid concrete panel. A common approach, instead of using a head flashing, is to install a two-stage sealant joint along the head and jambs of the windows and direct the water between the seals into the sill flashing for drainage. A proper seal of the top edge of the upturned end of the sill flashing to the concrete jamb is critical to prevent the water that flows down between the jamb seals from bypassing the sill flashing. We have used sheet rubber sill flashings adhered to the concrete and separate pieces of sheet metal, i.e., counterflashing, set into a sealant-filled reglet to cover the top edge of the upturned end of the flashing with success (Figure 13).

<u>Common problems</u>: Exposed aggregate finishes on the panels present an irregular surface for sealant adhesion (Figure 14). It is nearly impossible to tool the sealant into the surface irregularities, resulting in pinholes and leakage. A better approach is to use panels where the sides and perimeter of the face of the panel

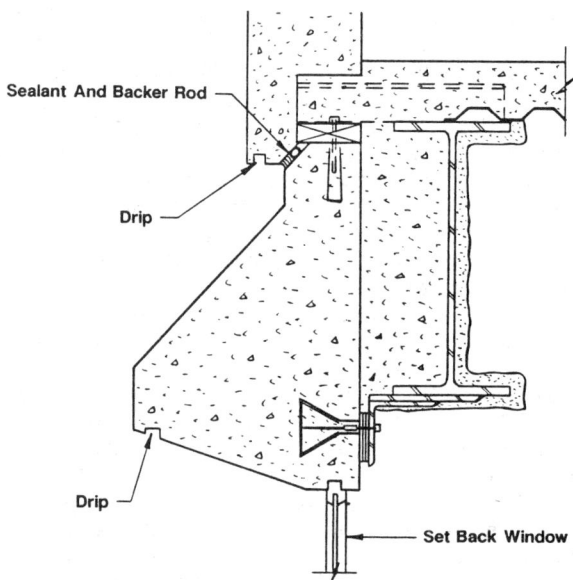

FIGURE 9. Vertical section showing horizontal joinery in precast concrete panels. Note ship-lap geometry and recessed sealant to shield the joint from the weather.

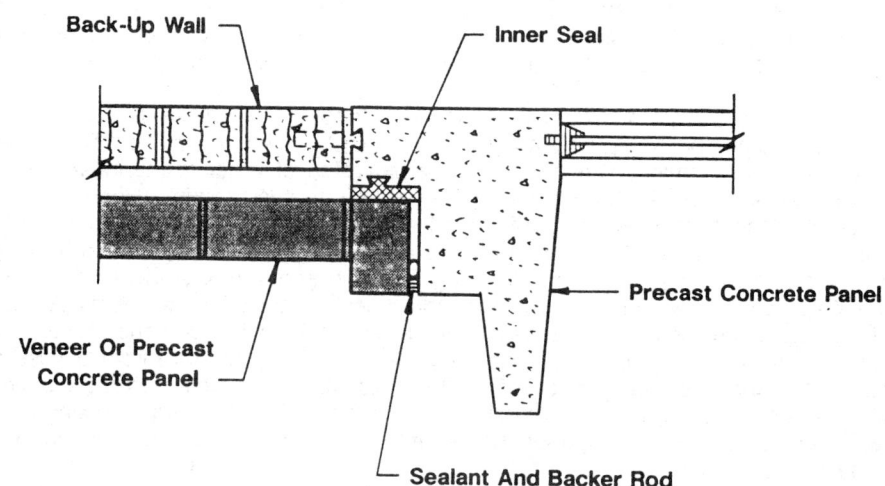

FIGURE 10. Plan section showing vertical joinery in precast concrete panels. Note that panel geometry shields vertical joint from weather. Water that penetrates outer seal does not have a direct path to the interior.

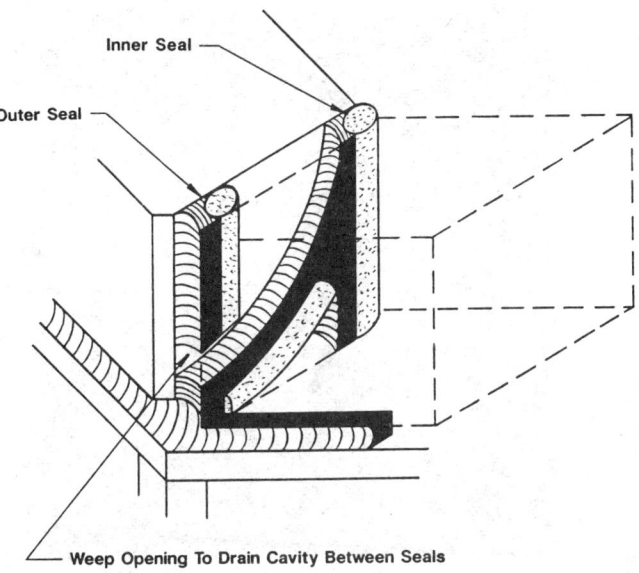

FIGURE 11. A two-stage remedial vertical seal with weep openings at the base of the joint.

FIGURE 12. Stainless steel flashing (A) installed above strip windows along the base of travertine-faced (B) precast concrete panels (C). The flashing turns up behind the concrete panels.

FIGURE 13. Mock-up of a flashing corner detail on a concrete wall. Note metal counterflashing (A) and sealant-filled reglet (B) to integrate the end of the flashing with the concrete.

FIGURE 14. A sealant joint along the edge of a precast concrete panel with an exposed aggregate finish.

is finished smooth and confine the exposed aggregate to the central portion of the panel.

Panels commonly develop hairline shrinkage cracks, particularly at the perimeter edges and at corners of punched openings, despite controlled curing procedures. These cracks create avenues for water to bypass shallow joint sealants, even when they have good adhesion to the panels (Figure 15).

Remedial options: Cracks through the panel can be epoxy-injected to prevent water penetration, provided the crack arose from overstress during improper handling and overstresses will not reoccur, such as with thermal bowing conditions. Access to both faces of the panel is required to construct a dam to retain the epoxy on one face and to inject the epoxy on the opposite face. The repair may blend well with the concrete when dry, but can stand out when the panel is wetted, due to the differences in porosity between the epoxy and the surrounding concrete.

Cracks in areas of ongoing movement require less rigid repair materials to maintain a seal and allow movement. The crack can be routed to form a shallow, narrow groove on the face of the panel, at least 9 cm x 9 cm (3/8 in. x 3/8 in.), release tape applied to the base of the groove, and liquid-applied sealant installed. The release tape is needed to distribute the crack movement over an unbonded area of the sealant and avoid strain concentrations in the sealant. This type of repair may not match the appearance of the surrounding concrete.

Generally, joint repairs involve upgrading single-stage sealant joints to two-stage seals that are drained with weep holes through the exterior seal. This can be difficult when existing joints are narrow, since access is needed through the joint near the back of the panel to construct the inner seal. Cutting the joint wider for some partial depth of the panel can resolve this problem, but may be costly.

Depending upon the configuration of the panels and their layout, it may be possible to install flashings along the base of the panels. With certain panel layouts, we have been able to slide a flashing into a horizontal joint between panels, although this is tedious and many obstructions such as anchors and shims arise. This type of repair is not used commonly.

Where flashings have been omitted from window sills, it is possible to install such flashings. In some cases, flashings can be installed without removing the existing windows, but often window removal is necessary and less costly. If windows are not removed, any existing sill anchors need to be cut and shims have to be removed. This approach also requires substantial clearance between the frame and the supporting structure, and is not feasible when narrow sill perimeter sealant joints exist. In some cases, it is possible to remove wood blocking below the window to increase the clearance. Flexible flashing materials, such as sheet rubber, are useful, since they can be slid through narrow openings and turned up on the inside of the frame.

Glass/Metal Curtain Walls

Curtain walls are metal-framed walls with various infill materials, glass being the most common. Most frames are assembled from individual horizontal and vertical members, i.e., stick systems. Metal-to-metal framing intersections and

30 WATER IN EXTERIOR BUILDING WALLS

FIGURE 15. An example of hairline cracks in the edge of a precast concrete panel along a window head perimeter sealant joint that can allow water to bypass the sealant.

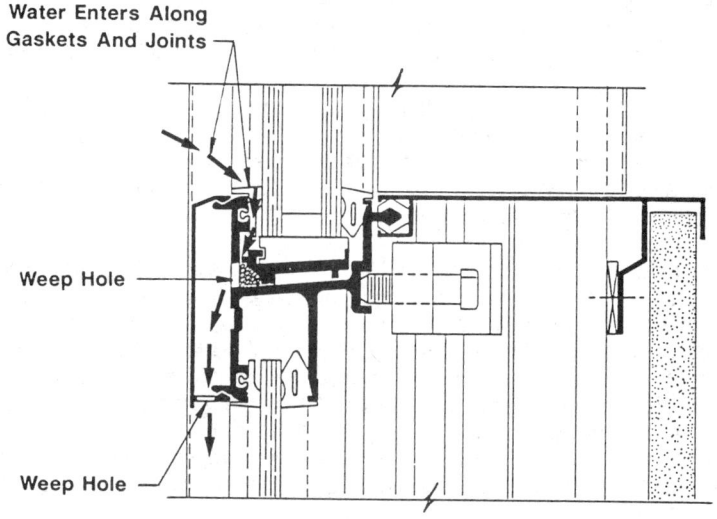

FIGURE 16. Cross section of a curtain wall. Note secondary drainage capability of pocket below glass. Base of pocket is sloped outward to promote drainage through weep holes.

glazing-to-framing joints are sealed commonly with foam or dense rubber gaskets, or liquid-applied sealants. At some corners, rubber plugs or pieces of metal are incorporated with the sealant to fill gaps in the framing.

These systems use a variety of cavity wall and pressure-equalized design principles for waterproofing. The manufacturers of these systems generally recognize that some water will penetrate the joints in the system, including the glazing and framing joint seals. Therefore, curtain walls generally are designed to drain the water that penetrates these joints down to each sill (horizontal framing member) where it is weeped back to the outside (Figure 16).

The sill member generally acts as a trough or gutter and, as such, it must have corners and intersections sealed permanently watertight (Figure 17). Weep holes in the sill should be protected by covers or shielded to avoid direct inward flow of water. In many systems, the inboard side of the sill gutter is sealed airtight, resembling to some degree pressure-equalized design.

Traditionally, curtain walls do not incorporate through-wall flashings, except at the base of the wall, relying instead on the sill gutters and their corner seals at vertical members to collect and contain penetrating water at frequent vertical intervals until it weeps out of the sill gutter. This approach lacks the reliability that a separate flashing provides, in that the corner seals are formed from liquid-applied sealants which are not as reliable or durable as the soldered corners in sheet metal flashings for example. Unfortunately, the service life of these sealants is much shorter than that expected for the wall system. While the system may perform well for many years, there typically is not a reliable means for replacing the seals in the future when they deteriorate. While this is a weakness, the systems generally perform better than barrier wall systems because they provide some secondary drainage capability and do not rely solely on a single exterior seal for waterproofing.

The critical requirement for these systems is providing a durable seal for the corner joinery where the horizontal member abuts the continuous vertical member (Figure 17). The most common means for creating this seal is to install pre-formed gaskets and/or liquid-applied sealant over the joined metal extrusions. The better designs incorporate the following features:

- A slight outward slope along the bottom of the horizontal member directs water outward and reduces the magnitude and duration of water contact with these critical joint seals (Figure 16). Prompt drainage limits leakage volume at any seal defects and improves the durability of the sealant which degrades when immersed in water.

- Frame extrusions without complex geometries and differing materials, such as screw bosses, offsets and thermal breaks, increase the chances of creating a continuous seal at corner intersections.

- Systems that permit construction of the corner seals in the factory, as opposed to on-site, generally have a greater chance of success due to better control on surface preparation and cleaning and better supervision of the sealing process.

Expansion joints in vertical members are a particularly difficult area in the framing to maintain a watertight seal. Generally, the joint incorporates a back-up plate behind gapped ends of the members, and the plate is bedded in a non-curing

FIGURE 17. View of corner detail of a curtain wall with the spandrel glass removed. Board insulation remains at right side of photo. Note seals around angle (A) and screw tips (B) to form sill gutter.

FIGURE 18. An expansion joint in a vertical mullion of a curtain wall. Note the gap in the seal created by the concentrated movement of the end of the upper member.

butyl-based sealant (Figure 18). One of the edges in such a joint faces against the flow of water, inviting water entry. Incorporating a back-up plate that fits behind the upper member and laps over the lower member is one approach to avoid this weakness, but it is not commonly used. With either approach, the movement of the vertical members concentrates at one point along the glazing seal and inevitably creates an unsealed opening at this point (Figure 18). Another alternative is to create a butt sealant joint at the expansion joint to avoid impeding the water flow and distribute the movement of the ends of the vertical members. This requires providing solid watertight end caps on the members, and is not commonly done. Typically, the systems accept these weaknesses and attempt to collect water that penetrates the expansion joints in the drained sills.

Common problems: The most prevalent problem that we find is defects in the corner joinery seals, including omission of the seals altogether due to fabrication and erection oversights in some cases. Other defects include the following:

- pinholes or discontinuities in the internal corner seals due to the complexity of the intersecting members and difficultly in accessing certain spots along the joint, and

- poor adhesion of sealants due to improper cleaning and surface preparation, incompatibility with various materials particularly at plastic and rubber components such as thermal breaks, and deterioration of the sealant material with age and exposure to ponding water.

Frequently, we find other defects in the external seals that allow significant water entry into the system, which tends to exacerbate leakage at any corner seal defects. Glazing seals are the most common source of water entry to the framing system. Gasket shrinkage or improper installation cause gaskets to pull away from the frame at the glazing corners (Figure 19). Gaskets require proper installation methods to avoid stretching and the resulting "shrinkage" over time as the gasket relieves this built-in stretch. Some gaskets also shrink due to material behavior, i.e., weathering and loss of plasticizers. Non-uniform compression on the gaskets, due accumulated fabrication tolerances or variable tightening of pressure bar glazing bead fasteners, can allow water penetration.

We generally recommend constructing external glazing seals with liquid-applied sealants, i.e., wet seals, as opposed to pre-formed dry gaskets. The wet seals avoid gasket joinery and compression pressure problems, since they are continuous and adhered to the substrate. Wet seals are subject to some defects due to installation tolerances, but our experience is that they prevent more water penetration than dry gasketed systems. They require outside access for glass installation and replacement, and sometimes are not used for this reason.

In some cases, such as at the base of a curtain wall or at windows, the sill frames are anchored to the structure with fasteners that penetrate the sill gutter and any underlying flashing (Figure 20). Fastener holes provide an avenue for water penetration. Sealant materials, if any, used to cover the fastener heads only provide short-term protection as they often loose bond when subject to "ponding" water. Figure 2 shows a detail for fastening the sill frame with a clip angle and fastener into the rear of the frame to avoid penetrating the horizontal portion of the sill flashing. Penetrations through the upturned rear leg have a very minor exposure to water, compared to those in the horizontal part of the flashing.

34   WATER IN EXTERIOR BUILDING WALLS

FIGURE 19.   A gap at the end of the rubber glazing gasket seal at the corner of a window.

FIGURE 20.   The sill of a strip window with the glass removed. Note the unsealed screw penetrating the bottom of the sill framing.

Remedial options: Remedies for leaking glass/metal curtain walls include two approaches using sealant materials; flashing installation generally is not feasible with these systems.

One option is to wet seal all external joints in the system, i.e., seal them with liquid-applied sealants. This option typically is tried because of it relatively low cost and low disruption. This approach does not treat the common fundamental defects that exist in the waterproofing system, i.e., leaking internal seals, but instead attempts to make a barrier system out of the curtain wall and prevent water from reaching the corner seals. As such, it contains the drawbacks of any barrier system, and some degree of on-going leakage is likely with the extent depending on the quality, durability, and maintenance of the wet seals.

Another option is to reconstruct the corner seals. This approach has included various schemes ranging from drilling portholes and blindly pumping sealant into the hidden corner areas, to partial disassembly of the curtain wall, including removal of glass lites, pressure bars, or frame members to repair the corner seals. We have never found the former approach effective due to the inability to clean, prepare and inspect the joint, while the latter approach can be successful, provided reasonable access to the joint for cleaning and remedial sealant application and tooling can be obtained. Wet sealing the system after corner seal repairs is prudent to reduce reliance on the internal remedial seals.

Exterior Insulation And Finish Systems

Exterior Insulation and Finish Systems (EIFS) typically consist of polystyrene insulation boards, which are covered by a polymer modified cementitious coating (synthetic stucco) that is reinforced with glass fiber mesh. Generally, the coating consists of two layers, a base coat and a finish coat, which is called the lamina. The insulation boards are usually adhered to exterior gypsum sheathing on a steel stud back-up wall; in some systems, the insulation boards are fastened mechanically to the steel studs. Most EIFS installations have been field constructed, as opposed to panelized, and have been adhered, rather than mechanically-attached, to the back-up wall.

EIFS systems use barrier wall principles and lack any cavity or waterproofed back-up. Traditional cement plaster stucco wall systems can incorporate a drainage layer through the use of asphalt-impregnated felt behind the metal lath. While this is not a clear drainage cavity, our experience is that the felt can control water that may penetrate at cracks or joints in the stucco wall, if it directs this water onto a through-wall flashing. However, this places the metal lath and fasteners in a moist environment and invites corrosion problems. With the EIFS composite of materials, such a waterproofing layer cannot be incorporated, because it would interrupt the adhesive attachment of the insulation or plaster coats. It may be possible to incorporate a waterproofing layer if the system is mechanically attached, but we have not seen such an approach in practice. Like traditional stucco, the fasteners with mechanically-attached EIFS systems are in a corrosive environment and subject to premature failure.

The EIFS systems rely solely on the polymer modified stucco coating and joint sealants to resist water leakage. Rain penetration through EIFS clad walls typically occurs at cracks in the lamina, at defects in the joint seals, and through unflashed window frame corners and joinery. Unlike most barrier wall systems, some components of the EIFS system, such as the gypsum sheathing to which it is adhered, degrade readily when exposed to water. Structural deterioration of the gypsum sheathing, fasteners and steel studs, and loss of attachment, become a greater concern than just discomfort of the building occupants and damage to interior finishes due to water leakage.

Control of cracking is important in these systems, particularly the control of cracks that occur over the joints between insulation boards. Hairline cracks that do not penetrate through the lamina have no leakage-related consequence. However, if cracks occur through the lamina, especially over joints in the insulation boards, water has a ready path to the water-sensitive exterior gypsum sheathing board, particularly under differential air pressures across the wall. Causes of cracking are discussed further in the section below.

Methods of waterproofing the joints between panels and the need for sill flashings at windows and other wall penetrations are similar to that discussed previously for precast concrete panels.

Common problems: Problems with EIFS systems can result from cracking of the lamina, which must remain unbroken for watertightness.

We have seen buildings where vertical or horizontal control joints are omitted and this has produced significant cracking, particularly on elevations with strong solar exposures (Figure 21). Some manufacturers of adhered systems have asserted that the system is "soft" and can "float" in response to thermal cycles. Consequently, these systems sometimes are designed without vertical control joints to subdivide building elevations into discrete panels. The polystyrene insulation has a relatively high coefficient of thermal movement. The lamina and the composite EIFS system have a lower coefficient, based on our testing of laboratory samples and measurement of movements on actual building walls, but the coefficient is sizable and requires due consideration in design.

We recommend that wall elevations be subdivided by control joints. Until designers can agree on a minimum joint spacing, we recommend spacing the control joints approximately 7 to 10 m o.c. (20 to 30 ft), since this generally is consistent with spacings used with other cladding materials. These control joints are in addition to those normally required by the manufacturer, such as at intersections of dissimilar materials or where structural movement may occur, i.e., vertical joints at intersecting walls and horizontal joints at floor levels with flexible edge beams or slabs.

Cracks typically develop at the re-entrant corners formed by window openings. In many cases, the insulation board joints align with the window corner creating a plane of weakness in the EIFS substrate aligned with a point of high stress caused by the window opening penetrating through the face of the panel. These cracks can allow direct water entry or water can bypass the window perimeter sealant where the crack and sealant intersect. Corner cracking can be reduced by cutting a single insulation board to fit each window corner such that board joints do not align with the window corners and by following manufacturers' recommendations to install extra layers of diagonally oriented reinforcing at all opening corners.

We have found that prolonged exposure to moisture softens some finish coats. At sealed panel joints, the softening can permit cohesive failure within the lamina when the joints move and the sealant pulls on the finish coat. Some manufacturers currently recommend omitting the finish coat from areas that receive sealant in

FIGURE 21. A crack in the EIFS coating below the sill corner of a window on a wall without any vertical control joints.

FIGURE 22. The EIFS coating and insulation has been removed at an unflashed, leaking sill corner of a horizontal sliding window. Note the debonding of the paper facer and the degradation of the core of the gypsum sheathing.

joints. Using low modulus urethane or silicone sealants helps reduce the stresses on the finish coat, but they do not have an extensive track-record of use in these systems.

During our field investigations and watertests, we have found that leakage from sill-to-jamb window frame corners penetrates behind the lamina and insulation when sill flashings are omitted from the window opening. As a result, the exterior gypsum sheathing often has significant hidden deterioration in the vicinity of such window sill corners (Figure 22). Horizontal sliding windows, which are commonly used in residential complexes, are particularly prone to frame corner leakage. The weatherstripping seals on the sliding joints tend to allow more water entry into the window system, especially as the weatherstripping deteriorates from use, than do seals on other styles of windows. In addition, the sill acts like a gutter as it does in a curtain wall, increasing the exposure of the corner joinery seals to water compared to other styles of windows where water does not collect in the sill.

In many of our leakage investigations, we find that these system problems are exacerbated by the flush-glazed, flat surface profile of the facade that does not shield the vulnerable surface seals.

Remedial options: Since EIFS is a barrier system with components readily damaged by water, the system requires frequent inspection and maintenance to limit water entry and consequential damages. Further, if significant leakage is occurring, a critical evaluation of the concealed conditions is needed to determine the scope of repairs.

Repair of cracks in the lamina vary with the cause of the crack. If cracks result from movements within the system that apply concentrated stresses to the lamina, remedial control joints should be installed to accommodate the movements. This requires removing the EIFS to form a joint and grinding the adjacent finish coat back to the existing base coat, wrapping the joint edges with reinforcing mesh and base coat that extends onto the back of the insulation (back-wrapping), and sealing the joint. To patch non-moving cracks, grind the finish coat back to the base coat and rout the crack. Fill the routed area with new insulation and rasp flush. Install new mesh reinforcing in new base coat, and a new finish coat.

With these systems, even simply cutting out the old sealant to repair defects can be a significant undertaking. Grinding to remove all traces of the old sealant, which is generally good practice when resealing, may damage the lamina. Bonding the new sealant to the remnants of the old failed sealant is not generally good practice, depending upon the materials involved. Upgrading single-stage sealant joints to two-stage joints is more difficult than with other wall systems, since the insulation and properly-applied coating may not extend deep enough to permit proper installation of dual seals.

Significant cutting and patching would be needed to install a remedial flashing to drain the joints in this system.

CONCLUSIONS

Exterior wall systems that incorporate cavity wall waterproofing principles are the most reliable in preventing water leakage to the building interior. The key component for these systems is the through-wall flashing which should be durable and have an expected service life equivalent to that of the entire wall system. Proper attention to the detailing and installation of these flashings is crucial to the success of a cavity wall system. Lower durability flashings with limited track records should be avoided due to the high cost of future replacement of failed flashings.

Barrier wall systems with modifications to incorporate some degree of secondary drainage capability, particularly at vulnerable joints, can provide levels of watertightness acceptable to some building owners, if sound, durable materials are used to form the barrier. Barrier walls that rely solely on surface seals and which use components that deteriorate readily from water that penetrates flaws in those seals do not provide a level of waterproofing reliability acceptable to most building owners.

All wall systems, and in particular barrier walls, can benefit from shielding provided by proper articulation of the wall surface to promote water drainage away from vulnerable joints.

Ultimately, the building owner should make an informed decision when selecting the wall system, based on the design professional's analysis of waterproofing reliability and the contractor's estimate of costs, i.e., affordability. Critical to the owner's evaluation is a clear understanding of the likelihood of leakage, the consequential damages from leakage, and life cycle costs and disruption associated with repairs, maintenance and replacement of the various cladding systems.

ACKNOWLEDGMENTS

We thank the principals and associates of Simpson Gumpertz & Heger Inc. for their support in writing this paper.

REFERENCES

[1] Design Windloads For Buildings And Boundary Layer Wind Tunnel Testing, AAMA Aluminum Curtain Wall Series No. 11, American Architectural Manufacturers Association, Des Plaines, IL, 1985.

[2] Myers, J. C., "Technics Topics - Window Sill Flashings: The Why and How," Progressive Architecture, June, 1990.

[3] "Water Resistance of Brick Masonry, Construction and Workmanship, Part 3 of 3," Technical Notes on Brick Construction 7B Revised, Brick Institute of America, Reston, VA, April 1985.

Robert J. Kudder and Kenneth M. Lies[1]

DIAGNOSING WINDOW AND CURTAIN WALL LEAKS

REFERENCE: Kudder, R. J., and Lies, K. M., "Diagnosing Window and Curtain Wall Leaks", <u>Water in Exterior Building Walls: Problems and Solutions, ASTM STP 1107</u>, Thomas A. Schwartz, editor, American Society for Testing and Materials, Philadelphia, 1991.

ABSTRACT: Determining the cause or causes of leaks, isolating the problem components and details, and verifying the efficacy of remedial measures, requires systematic diagnostic procedures. Documentation of the damage, evaluation of project documents, determination of as-built conditions, and testing, are necessary steps in a comprehensive program. Discussed are the evaluation and testing procedures, masking techniques, disassembly and inspection procedures, etc., which have proven useful as part of solving leakage problems. Particular emphasis is placed on testing procedures which are adaptable to use on-site.

KEYWORDS: curtain walls, inspection, leaks, testing, water infiltration, windows.

Our industry knows what has to be done to avoid leaks. In the design phase of a project, attention to product and material selection, careful detailing, meaningful performance specifications and requirements for testing, anticipation of construction sequence and interfacing problems, and thoughtful shop drawing reviews are essential. In the construction phase, maintenance of reasonable tolerances, careful handling and installation, adherence to manufacturer's recommendations and project documents, and skillful execution are essential. Omission of any one of these elements is an invitation for leaks. Unfortunately, our industry also knows that it is possible to do all the right things and still create a building that leaks. Leaks occur even when well intentioned and skilled people are involved in a project.

Seeing a leak or water damage, and responding appropriately, can be thought of as a reprieve from even greater problems in the future. Imagine the deterioration and damage possible from a concealed leak which remains undiscovered or which does not cause visible damage, or from an observed leak which is ignored. When leaks occur, it is essential to avoid a stampede to "do something right now". Posturing, uninformed finger

---

[1]Dr. Kudder is a Principal, and Mr. Lies is a Project Architect II, with Raths, Raths & Johnson, Inc., 835 Midway Drive, Willowbrook, IL 60521.

pointing, blasting with water from fire hoses, and the automatic response of "more sealant" do not identify and fix leaks, and often make an eventual cure more difficult. A systematic approach to diagnosis and repair in an atmosphere of cooperation among all parties involved is the most expedient and economical path to a solution.

In the process of evaluating leaks, the writers have encountered recurring causes which could have been avoided, as well as recurring causes which can be exposed by on-site diagnoses. The review, inspection and testing procedures which have proven useful in identifying and understanding leaks are discussed.

DIAGNOSTIC PROCEDURES

A systematic diagnosis does not necessarily begin with testing. Three things should be understood before testing begins: the extent of the problem; the design intent; and the as-built conditions. On-site testing will efficiently yield useful information only after these three aspects of the wall are considered.

Understanding the Problem

Gathering information on the extent of the leakage problem serves two purposes. First, patterns in the observed leakage and damage often provide clues as to the cause. More importantly, this information provides a checklist against which conclusions can be evaluated. A comprehensive diagnostic program must lead to an explanation of all aspects of the observed leaks and damage. Conversely, if only a portion can be explained, the program has not been comprehensive and some leak sources have probably remained unidentified. Unfortunately, the corollary is not true. Being able to explain all aspects of the observed damage does not automatically mean that all leak sources have been identified. Multiple sources can combine to create a single symptom and a single source can create multiple symptoms. Leakage problems are not always attributed to window and curtain wall water infiltration. Other sources such as plumbing problems and condensation potential must be considered. The more complete the information regarding the extent of the leaks, the easier it is to resolve these difficulties.

Asking questions and systematically recording observed damage on drawings are the tools of information gathering. There are several questions which are useful to ask. The answers can often be obtained from subcontractors, tradesmen, occupants, maintenance personnel or first hand observations. The questions which should be asked include the following:

1.1 When a leak occurs, where does it appear to come from?
1.2 Under what conditions does the leak occur?
    a. Was the leak a one-time occurrence under exceptional conditions, or is it a recurring problem?
    b. Does it occur only when it is raining, or does it occur only in cold weather with or without accompanying rain?
    c. Does it start immediately after rain begins or sometime later?
    d. Does it end when the rain ends, or does it linger on for some time after?
    e. Does it occur during every rain or only when the wind blows from certain directions or with a certain severity?
    f. Does it occur only under certain conditions of building operation?

1.3 Upon review of a tabulation or recording drawing showing all known observations of leaks and damage, such as the drawing in Figure 1., is there a pattern?
   a. Do they occur only in certain portions of the building which have an exterior feature in common?
   b. Do they it occur only at certain elevations that have a unique feature such as exposed balconies, roofs or canopies overhead?
   c. Do they occur in a consistent location relative to some feature of the wall?
1.4 What steps have already been taken to deal with the leakage problem?
1.5 If attempts at a repair have already been made, are there any areas of the building still in their original condition?

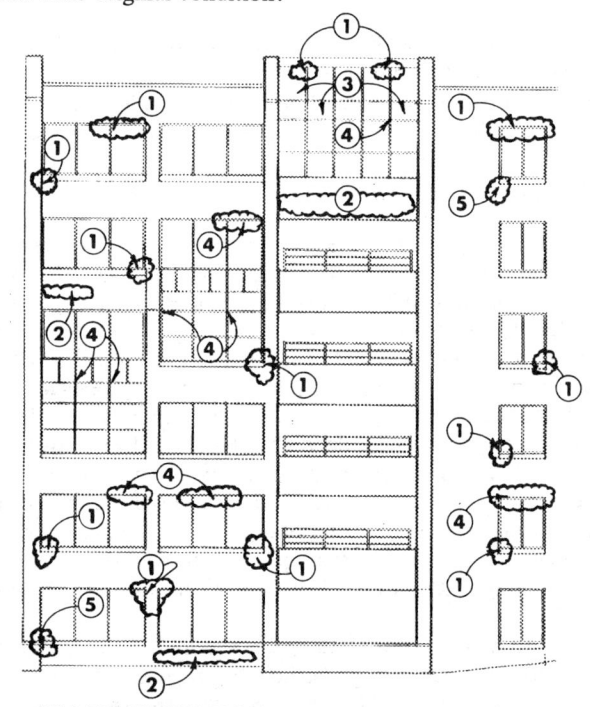

INTERIOR CONDITIONS

1. Stains and Deteriorated Finishes
2. Wet and Stained Carpeting
3. Failed Insulated Glass
4. Leakage Observed
5. Cracked Wallboard

FRONT ELEVATION

FIG. 1 -- Overlay drawing recording interior damage on an exterior elevation.

Understanding the Design Intent

A review of the project documents is necessary to understand the design intent, including how the system is supposed to function and what the required performance levels are. It is important to obtain a complete set of project documents for review. In addition to the project specifications and architectural drawings, other key documents include: addenda, bulletins, project correspondence, installation instructions, and approved shop drawings. Curtain wall and window leakage can be a direct result of design and specification deficiencies, or differences between the design intent and the as-built condition. The primary objective of the design review is to gain familiarity with the design and to check key elements and requirements of the specifications and detailing.

Specifications--The project specifications provide information on window or curtain wall performance requirements. The specifications should be clear, concise and free of ambiguity. A loosely written specification with inconsistencies may be interpreted contrary to the design intent or project requirements. Unclear specifications or inappropriate product selections can be a factor for curtain wall and window non-performance.

For window products, ANSI/AAMA 101-88[1] is an industry guide specification; requirements for various product types are standardized by using classifying designations which indicate the product configuration, grade and performance class. While the standard designations are suitable for most projects, the designer must consider customizing the performance criteria to meet specific project requirements. AAMA GS-001[2] provides for the special requirements of a specific project. Architectural windows are generally considered for use in heavy commercial or monumental buildings that require higher than normal performance. The building's occupancy is also a factor in properly selecting a window product. Building types, such as libraries and hospitals, operate under high interior humidity levels which can lead to severe condensation problems if the condensation resistance factor (CRF) is improperly specified. Geographic location is another important consideration when specifying performance criteria. Buildings located in regions that are subject to simultaneous peak wind pressures and rain, such as gales and hurricanes, require greater water infiltration performance requirements.

The air infiltration, water infiltration and structural performance criteria given in a specification are meant to be demonstrated by laboratory tests. Tests are used to qualify a product for inclusion in a project by verifying compliance with the specification or as a method of quality assurance of production units. The product classifying designations provide indications of a component window's performance rating at a specific size. Additional laboratory testing may be necessary to determine the performance of oversized units, configurations that involve stacking, or units that combine numerous components into one assembly. Full-scale wall mock-up tests are commonly required for custom curtain walls, especially if the geometry or material application is unique. Air infiltration, static and dynamic water infiltration, and structural tests are usually performed. Mock-up drawings prepared for these tests should indicate any changes or alterations made as a result of the testing.

The specifications should clearly define the water infiltration test failure criteria. Opinions can differ considerably between what is regarded as acceptable performance. The specifications should address:

2.1 Where water may enter and where it may not enter?
2.2 Where temporarily retained water may accumulate and how much may accumulate?
2.3 Which components are included in the water control system for the purpose of a product test?
2.4 Which components are secondary defense features of the architectural design, such as flashings, and should they be considered as part of the window or curtain wall drainage system for the purpose of performance certification?

To the extent possible, responsibilities for the installation and performance of windows and curtain walls should be made clear in the contracts. If the responsibility can be assigned to one or two parties, there should be little question about who must participate in solving a leakage problem.

Drawings--The integration and detailing of all curtain wall or window components and materials are shown on the architectural design drawings and the shop drawings. Differences between the design and shop drawings should be identified. Often the reason for the differences is not readily apparent, and may result in speculation and ambiguities. Therefore, records should be kept throughout the project's development which include a history of pre-qualification and laboratory testing results, modifications made in order to pass the tests, and the reasons for modifications.

The architectural design drawings may be very schematic or ambiguous about the integration and detailing of the wall components. This expedient unwisely places much of the design and detailing responsibility with the manufacturer or fabricator producing the shop drawings. For a system with a single major supplier, this responsibility can be accepted and addressed properly. For a system with a variety of components and suppliers, relinquishing design responsibility increases the likelihood of poorly conceived details and a fragmented approach to the integration of components. It also reduces the usefulness of the design drawings. During the diagnostic process, difficulties in understanding the design intent because of incomplete information in the design documents are the same difficulties that the producers, fabricators and erectors experienced during construction. Appreciating these difficulties can be a useful tool in understanding the problem.

The review of project documents should lead to clear answers to the following questions:

3.1 How was the wall intended to function? Is the overall design concept:
    a. A barrier wall in which all water infiltration paths are sealed?
    b. A drainage system incorporating a means of collecting and discharging infiltrating water to the exterior?
    c. A rain screen system which controls infiltration by pressure equalization?

3.2 Are the drainage paths and air seals clearly shown?

3.3 Do components have sufficient height to resist water lifted by differential air pressure or are these infiltration paths sealed?

3.4 How are the pieces of the system to be assembled? Is there an isometric drawing to clarify the fit-up of connecting pieces?

3.5 What is supposed to be done at atypical locations, such as:
    a. At the corners and terminations of flashing and receivers?
    b. At splice joints and expansion joints?
    c. Where two different cladding systems interface?
    d. At the top and bottom of the system?
    e. At beam-column intersections, slab projections, and other locations where the curtain wall may interface with the structural frame?

3.6 Are all sealants properly specified, configured and detailed?
    a. Is there sufficient volume for the expected movement?
    b. Has three-sided adhesion been avoided and do all moving joints include a bond breaker or backer rod?
    c. Are bonding surfaces sufficiently wide or is sealant supposed to adhere to sawed or extruded edges?
    d. If a backer rod is required, is the geometry of the joint adequate to confine the rod and hold it in place.?
    e. Has the appropriate type of sealant been selected for each application?
    f. Are the specified sealants compatible with the substrate and other materials they will contact?
    g. Are sealants susceptible to UV degradation properly sheltered?

h. Are substrate preparation and cold weather conditions addressed?
3.7 Where elements overlap, has the thickness of the materials been accounted for in the details?
3.8 Can the details actually be built?
   a. Are tolerance requirements reasonable?
   b. Will the completion of one wall component interfere with access for the completion of another component?
   c. Does more than one component have to occupy the same place, or do mutually exclusive activities have to occur simultaneously?

Questions of this type which can not be satisfactorily answered during a review of the project documents because of ambiguities or a lack of information certainly suggest the need for further investigation.

Understanding the As-Built Condition

The actual as-built conditions of the subject building also must be understood. This can only be accomplished by thorough inspection. The inspection process associated with the diagnosis of leaks may be the first and only time that a wall is considered in its entirety. The components of the wall, the structural elements supporting the wall, the utilities within the wall, the finishes on the wall, and the penetrations of the wall should work together. The drawings will rarely predict the relationship between all these aspects of the wall accurately. In the design process, attention is fragmented by discipline. In the construction process, activity is fragmented by trades. It is sometimes impossible to determine from the drawings how all of these wall features are intended to fit together.

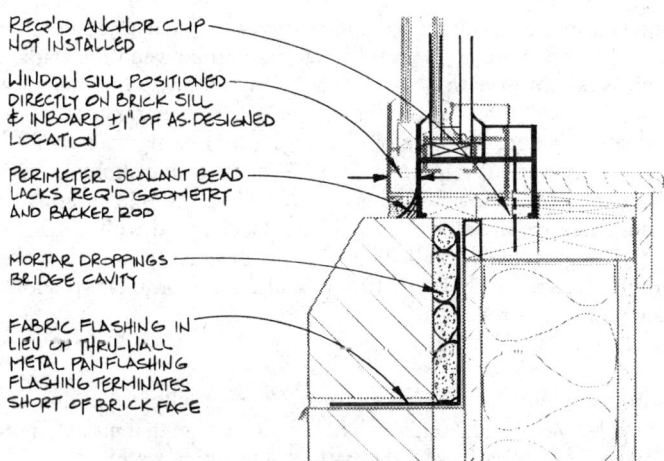

FIG. 2 -- Composite inspection drawing, with design information in half-tone, and inspection observations shown with solid lines.

Composite drawings are helpful for documenting as-built conditions. These drawings depict how the wall was intended to be constructed, based on the best available information found in the project documents. The writers have found that a half-tone copy of this drawing is an ideal media for recording information from inspections. The contrast between the half-tone image and the solid lines recorded during the inspection make comparisons between design and construction easier, as shown in Figure 2. It is important to remember that differences between design and construction do not always adversely effect the performance of a wall and may be for the better. Discovering a difference does not necessarily mean that a leak source has been identified.

FIG. 3 -- Observations of as-built conditions possible only by partial disassembly of the system.

A thorough inspection requires openings to reveal concealed conditions and selective disassembly of wall components to reveal the relationship between the parts, as shown in Figure 3. Openings do not have to be large. The use of an inspection mirror with a swivel head and extendible handle, and a flashlight, will extend the range of view at a small opening. For locations where patching of openings would be particularly difficult, a fiber optic borescope can be used. It requires only a 5/8 inch diameter hole for insertion of the shaft and provides a 360 degree view within the wall. Each step in creating an inspection opening or during selective disassembly should be documented with notes, sketches, and photographs. It is impossible to anticipate the significance of some observations until further demolition or disassembly is completed, and relying on memory alone to make a retroactive evaluation is not reliable.

The inspection process also provides an opportunity to evaluate the current condition of concealed gaskets and sealants, concealed damage that may be more significant than visible damage, and indications of the path of infiltrating water.

DIAGNOSTIC TESTING

Testing is the essential last step in most diagnostic programs. It will confirm a hypothesis derived from a design review and inspection, reveal leakage paths and resolve differences of opinion about causes, and complete the information necessary for a thorough evaluation. If a theory on the cause of a leak can not be demonstrated by a reasonable test, it is probably wrong.

Diagnostic vs. Qualifying Tests

Test methods designed to identify and/or quantify window and curtain wall leakage can be performed in a laboratory on a representative sample or at the building site on an installed unit. Laboratory tests and mock-up tests are usually performed to demonstrate that a window or curtain wall meets the performance requirements of the project specifications or as a method of quality assurance. Similar tests can be conducted to diagnose problems and to determine current performance.

Many factors can affect the current in-place performance of a window or curtain wall. The performance criteria provided in industry guideline specifications pertain to the unit as a product and not the overall installation, unless specifically stated so in the project specifications. These tests are performed under laboratory conditions. Normal wear and tear on the unit, abuse or damage, lack of maintenance, modifications, and aging of materials can cause a unit to perform differently than it might have in a laboratory when it was first manufactured and pre-qualified. In the evaluation of an installed unit that has been in service, it is necessary to distinguish these causes of non-performance from intrinsic properties which would cause non-performance of the unit, no matter when in its service life it was tested. Making this distinction is part of the responsibility of the engineer conducting diagnostic testing.

Selecting a Test Method

Diagnostic test methods and procedures must be responsive to both technical and non-technical constraints. Testing costs can vary significantly between various methods, and the project budget may be an important client consideration. The decision to remove a unit and test it in the laboratory, or to test on site, must be made. If extensive modifications and retesting of a representative unit are anticipated, there are advantages to removing one and testing it in a laboratory. To minimize racking and disturbance of a removed sample, it is prudent to secure diagonal braces before removal begins. If leakage at the interface between wall components is suspected, tests should be performed on site. Physical limitations of access at the site or a directive to minimize disruption must also be considered.

The objectives and failure criteria of the test should be clearly defined and understood. This preliminary step will help resolve conflict in the interpretation of the results. Common objectives include:

4.1 Determining the level of in-place performance.
4.2 Verifying compliance with the project specifications.
4.3 Determining the source and location of leakage.
4.4 Determining the cause(s) of non-performance.
4.5 Evaluating the consequences of non-performance.
4.6 Determining if recommended remedial measures are feasible.
4.7 Verifying effectiveness of remedial measures.

There are several standardized test methods and procedures that are designed to be used in diagnosing window and curtain wall leakage attributed to rain infiltration. Each test method has limitations. Modifications to standardized test methods or custom devices can also be effective in isolating a problem. However, it is advisable that the impact of such modifications be fully understood and properly identified. In general, a test which simulates the effect of wind driven rain requires a chamber for maintaining and controlling a differential air pressure across the test area, and a matrix of spray nozzles for depositing water in a uniform and controlled manner on the exterior surface.

Testing Techniques

A wall can be tested as one large entity or as a collection of individual components. It is usually prudent to do both. Masking is used to isolate individual wall components during a test, as shown in Figure 4. Inexpensive 6 mil plastic sheeting is taped on the exterior to protect selected components from the water spray. If the perimeter of the mask is well sealed, the protected component is essentially excluded from the spray. Leaks observed during a test can therefore not be attributed to infiltration through a masked component. By moving the mask and retesting, each component can be tested individually. As an example in window testing, masking of adjacent construction, including perimeter sealants, is done so that the performance of only the window can be evaluated. Test records should always indicate the boundaries of the masking, and whether it influenced the results of the test.

Standard test methods using static air pressure difference are described in ASTM E 1105[3] and AAMA 502-90[4]. Two different methods of executing this test are acceptable, but each has limitations. The basic difference between the methods is whether the pressure chamber is located on the exterior or interior.

FIG. 4 -- Masking adjacent masonry to isolate a window for testing.

The test unit is subjected to a uniform water spray from a calibrated spray rack. The volume of water is monitored by a flow meter. A sealed chamber is mounted to either side of the unit and a blower is used to create a differential pressure across the unit. The differential pressure is measured by a manometer. The combination of water spray and differential pressure simulates a wind-driven rain.

Interior chambers--For an interior installation, the chamber should be sealed to the inside face of the perimeter receptors if these extrusions are used. Otherwise, the chamber should be sealed to the inside face of the frame extrusions. A limitation of this method is that perimeter seals and pan flashings are not tested under pressure. Interior chambers can be constructed of collapsible, light framing members and clear plastic sheeting, as shown in Figure 5. Such chambers are relatively inexpensive to install and convenient to transport.

More expensive custom-built chambers may be required to accommodate full-story curtain wall tests.

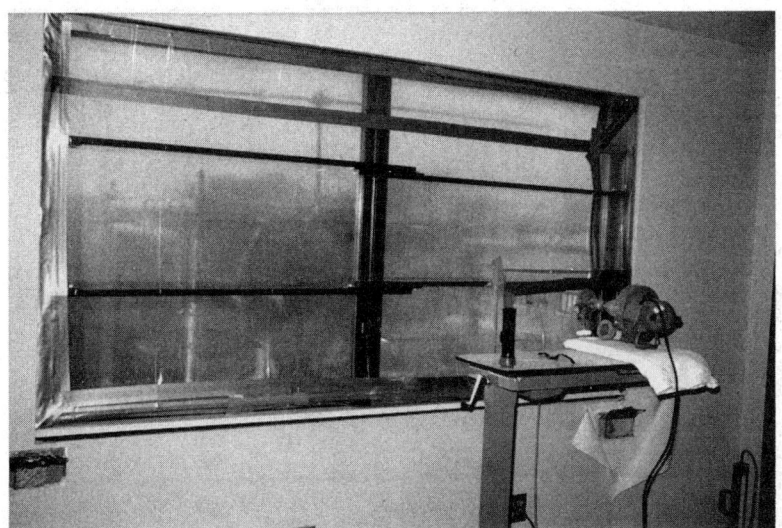

FIG. 5 -- Interior test chamber using telescoping lightweight metal frame.

The interior test chamber must not interfere with the response of the window and thereby invalidate the test result. The writers have observed tests conducted by other parties in which the plastic sheet that makes up the interior chamber is allowed to draw tightly against the window under negative chamber pressure. This eliminates the net air pressure on the glass because the chamber plastic sheet is reacting against it. Weather-stripping, glazing gaskets and tapes, interlocks, etc., are therefore not subjected to structural loads. The writers have also observed tests in which coffee cups, short lengths of plastic pipe, blocks of wood, or pieces of styrofoam insulation are propped between the chamber plastic sheet and the glass in the test unit. The plastic sheet is still reacting against the glass through the props. Tests conducted in this manner are meaningless and should not be accepted as a valid demonstration of performance.

Exterior chambers--Exterior chambers enable testing of all components and seals and do not necessarily have size limitations. Exterior chamber tests can also be less disruptive to the internal functions of an occupied building. However, because they are typically custom built and can be quite large, they are usually expensive and require more planning than interior chambers. Anchoring the chamber to continuous glass and aluminum curtain walls often requires an imaginative support scheme.

To facilitate masking, tare measurements, and trial modifications, an exterior chamber may have to be removed and reinstalled several times. An alternative is to design the chamber so that it can be entered from the outside and used as a work platform, as shown in Figure 6.

Simple preliminary tests--Several effective but simple methods are available for preliminary testing. The easiest is to check the watertightness of sill corners and sub-frame systems by flooding the track sections with water to create a static head. Plugging weep holes or creating a dam may be necessary to hold the water. Adding different water color dyes is also helpful in pin-pointing a leaky internal joint in a system with multi-chambered sills or stacked framing.

FIG. 6 -- Custom exterior chamber, which also serves as a work platform with access from the floor below.

FIG. 7 -- Using a smoke pencil to visualize air paths.

A smoke pencil device, shown in Figure 7, generates a small stream of white chemical smoke when a rubber bladder is squeezed. It makes it possible to identify air leakage paths. Commercial devices are marketed through mining supply companies. An air path in window and curtain wall framing may also be a water path or, in cases of condensation problems, an avenue for water vapor transport. Infiltrating air can cause percolation and spattering of water held in the system. The smoke pencil can also be used to find voids in testing chambers.

A calibrated hose nozzle, as shown in Figure 8, can be used to check curtain wall joints and perimeter seals for leaks. The test method is described in AAMA 501.2-83[5]. This test method is intended to verify water resistance of only joints and seals which are inoperable. The procedure involves systematically spraying approximately 5 feet of joint at a distance of 1 foot for a period of 5 minutes. The required 35 psi water pressure is monitored by a pressure gage and adjusted with a control valve. This test method is inexpensive and effective in isolating water entry points.

<u>Custom designed devices</u>--Custom designed devices for special applications can be used instead of a chamber and standard procedures to obtain useful information about specific water infiltration problems in a specific component. It is only necessary for the applied pressures to be controlled and reproducible and for the device to be compatible with the behavior of the component. The small device shown in Figure 9 was designed to test the watertightness of the edge band on a corrugated core metal panel system. The discharge tube on the right is a standpipe, which controls the differential pressure by the static head of water.

FIG. 8 -- Calibrated nozzle in use.

CONCLUSIONS

A comprehensive and systematic diagnoses of window and curtain wall leaks is a four step process:

5.1   Inspection and documentation to understand the problem.
5.2   Review of project drawings, specifications and shop drawings to understand the design intent.
5.3   Inspection and disassembly to understand the as-built conditions.
5.4   Testing to understand behavior, to verify theories about the causes of leaks, and to demonstrate the effectiveness of repairs.

Useful techniques are available for each step in the process. Some testing techniques are field adaptations of laboratory methods, while others are uniquely suited to field applications.

The diagnostic program must be comprehensive. Repair decisions will be based on the results of the program. The resulting repair recommendations will not solve the leakage problem if a source of leakage remains unidentified because of an incomplete diagnostic program.

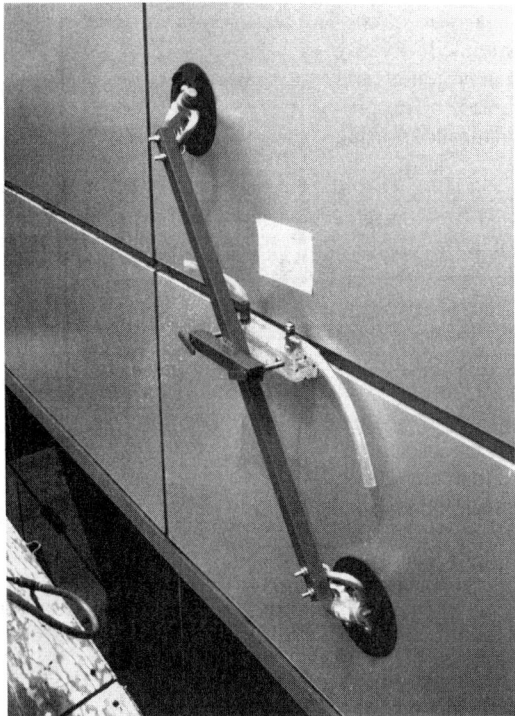

FIG. 9 -- Custom designed device for testing water infiltration at the edge band of a metal panel.

REFERENCES

[1]   ANSI/AAMA 101-88, "Voluntary Specification for Aluminum Prime Windows and Sliding Glass Doors", American Architectural Manufacturer's Association, Des Plaines, Illinois, 1988.

[2]   AAMA GS-001, "Voluntary Guide Specification for Aluminum Architectural Windows", American Architectural Manufacturer's Association, Des Plaines, Illinois, 1984.

[3]   ASTM 1105, "Field Determination of Water Penetration of Installed Exterior Windows, Curtain Walls and Doors by Uniform or Cyclic Static Air Pressure Difference", American Society for Testing and Materials, Philadelphia, Pennsylvania, 1988.

[4]   AAMA 502-90, "Voluntary Specification for Field Testing of Windows and Sliding Glass Doors", American Architectural Manufacturer's Association, Des Plaines, Illinois, 1990.

[5]   AAMA 501.2-83, "Field Check of Metal Curtain Walls for Water Leakage", in AAMA 501-83 Methods of Tests for Metal Curtain Walls, American Architectural Manufacturer's Association, Des Plaines, Illinois, 1983.

Robert G. Thomas, Jr.

WATER VAPOR BEHAVIOR IN EXTERIOR INSULATION AND FINISH SYSTEMS

REFERENCE: Thomas, R. G., Jr., "Water Vapor Transmission Behavior in Exterior Insulation and Finish Systems," <u>Water in Exterior Building Walls: Problems and Solutions, ASTM STP 1107</u>, Thomas A. Schwartz, Ed., American Society for Testing and Materials, Philadelphia, 1991.

ABSTRACT: Exterior Insulation and Finish Systems (EIFS's) are a type of cladding for exterior building walls. They are attached to the outside face of the wall structure. EIFS's were developed in Europe where exterior wall structures are predominantly unit masonry and concrete. In America, EIFS's are usually attached to lightweight wall structures of studs and gypsum sheathing.

In America, EIFS's are used in a wide range of climate conditions, including air conditioned buildings in hot, humid climates as well as arctic conditions. EIFS's are used in virtually every type of building including residential and high-rise commercial structures. Because of their light weight and relatively short track record, the longevity of EIFS's has been questioned. In particular, the effects of water from both liquid and water vapor sources is a concern. This concern has come from both field experience and theoretical analysis.

The American way of using EIFS's results in different water vapor behavior in the wall assembly. Condensation can occur in different locations in the wall under different conditions. The effects of water on the wall can be pronounced. The durability of the wall can be greatly affected if water vapor problems occur.

This paper reviews the American style of using EIFS's on light weight wall structures in terms of water vapor flow behavior. A computer program is used to demonstrate various conditions that affect the development of condensation in EIFS wall systems. The paper identifies common American wall designs that require attention and offers solutions for dealing with water vapor problems.

KEYWORDS: Water vapor transmission, condensation, moisture, dew point, insulation, Exterior Insulation and Finish System, vapor retarders.

Robert Thomas, Jr. is a principal of the construction products consulting firm CMD Associates, 21236 Tramp Harbor Road, Southwest, Vashon, WA 98070.

THE AMERICAN WAY OF USING EIFS'S

Description of an EIFS

An Exterior Insulation and Finish System (EIFS) is a type of wall cladding for exterior building walls. There are several types of EIFS's in the American construction market. This paper concerns Type PB EIFS's, as defined by the Exterior Insulation Manufacturer's Association. A typical Type PB EIFS is shown in Figure 1.

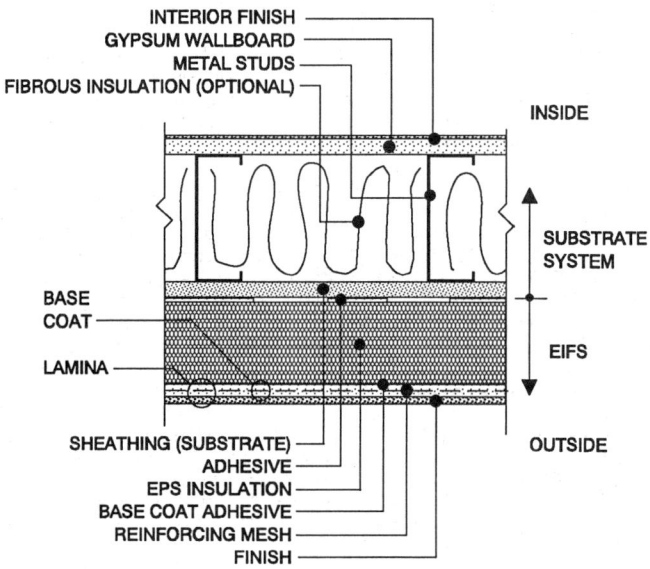

FIGURE 1. - Horizontal Section Through Typical EIFS

A Type PB EIFS consists of two basic layers. The first is a layer of Expanded Polystyrene Insulation that is adhesively bonded to the face of the wall. The second is a multi-layer reinforced, synthetic coating system that is applied to the insulation. EIFS's are usually installed directly onto the wall at the site, but can also be prefabricated as panels. EIFS's are usually installed by hand by plasterers. EIFS's are non-structural; they do not support the wall. EIFS's provide insulation, weathertightness and a finished exterior surface.

The exterior wall structure is called the Substrate System. The outermost surface of the Substrate System is called the Substrate. The type of Substrate System covered by this paper consists of wood or metal studs to which has been attached some type of sheathing board material.

A number of aspects of EIFS's affect their water vapor behavior, as follows:

Substrates: EIFS Substrates can be several materials. The most common is Exterior Grade Gypsum Sheathing. Other types include cement boards and plywood. Gypsum boards have limited inherent resistance to water. Thus, it is important that the water be kept away both during construction and when the building is completed. This includes water from leaks as well as from condensation. This is especially true with Type PB EIFS's because the only thing holding the EIFS to the wall structure is an adhesive.

Barrier Systems: EIFS's are barrier-type cladding systems. This means that they rely on the continuity of their outer surface to keep water from getting into the wall structure. This contrasts to drainage-type cladding systems that have been-in provisions for the removal of water. The fact the EIFS's are barrier-type systems is important from a water vapor standpoint because condensation within an EIFS has no path to the outside as a liquid.

Seamless: EIFS's have a seamless exterior surface. Type PB EIFS's do not require joints of themselves. Joints are required when the wall to which the EIFS is attached has them, where significant structural movement occurs, and where the EIFS starts and stops. This is important from a water vapor standpoint because the lack of joints in the surface means that an EIFS must allow the passage of water vapor by diffusion rather than by direct transfer through openings.

Joints: When joints do extend entirely through an EIFS, they are normally sealed by using a caulking-type material. Usually the back side of such joints is open to the stud cavity. This means that the water vapor pressure at the back side of the sealant joint is nearly the same as that of the stud cavity, but at a lower temperature. This is important from a water vapor standpoint because it can induce condensation at a point in the wall that is vulnerable to leakage.

Location of Insulation and Coatings: The external location of the insulation and coatings has important benefits from an energy standpoint. It also makes the temperature and water vapor profile of the wall different compared to traditional wall assemblies in which the insulation is inside the wall structure. This is important from a water vapor standpoint because the coatings must be breathable when placed on the outside. This allows the moisture flow to remain continuous through the outermost layers of the wall.

IDENTIFICATION OF PROBLEM AREAS IN EIFS WALLS

A number of EIFS wall design conditions have been identified that are sensitive to water vapor problems. These include:

- Condensation at the EIFS-to-Substrate bond area.
- Reverse vapor flow during winter versus summer.
- Use of fibrous insulation in the stud cavity.
- Placing coatings on the outside of the EIFS Lamina.
- Condensation in sealant joints.

These wall conditions are described in more detail on the following pages.

ANALYSIS METHOD

A method frequently used to analyze water vapor flow in building walls is the so-called "Steady State Method" described in the ASHRAE *Handbook of Fundamentals[1]*. A commercially-available water vapor transmission analysis program, *WVT[2]*, that uses the Steady State Method, is used in this paper to analyze water vapor flow in EIFS's.

The Steady State Method

The Steady State Method uses the following procedure:

- The wall is broken down into a series of layers.
- The thermal and vapor resistance of each layer is assigned.
- The indoor and outdoor temperature and relative humidity conditions are established.
- The temperature at each Interface is calculated.
- The saturation vapor pressure at the Interface temperature is calculated.
- The actual vapor pressure at the Interface is calculated.

If the actual vapor pressure at a given Interface is greater than the saturation vapor pressure at the same Interface, then condensation is likely to occur in the preceding layer.

Figure 2 shows a computer analysis of a wall using the Steady State Method. The following abbreviations are used:

MC	Material Code
INSIDE TEMP	Indoor Temperature (°F)
INSIDE RH	Indoor Relative Humidity (%)
OUTSIDE TEMP	Outdoor Temperature (°F)
OUTSIDE RH	Outdoor Relative Humidity (%)
L	Layer Identification
THK	Thickness of Layer (in.)
DESCRIPTION	Description of Layer
RT	Thermal Resistance (Btu/h/Sq. ft./°F)
RV	Vapor Resistance (Grain/h/Sq. ft./in. Hg.)
RT TOTAL	Total Thermal Resistance (Btu/h/Sq. ft./°F)
RV TOTAL	Total Vapor Resistance (Grain/h/Sq. ft./in. Hg.)
DIST TO X	Distance from Interface where condensation is first detected to point where it starts (in.)
TEMP AT X	Temperature at Point X (°F)
RATE AT *	Condensation at Interface where condensation is first detected (Grain/hr/Sq. ft.)
LL	Interface between Layers (AB is between Layer A and Layer B, etc.)
TEMP	Temperature at Interface (°F)
VPSAT	Saturation Vapor Pressure at Interface (in. Hg.)
VPACT	Actual Vapor Pressure at Interface (in. Hg.)

Limitations of the Steady State Method

The Steady State Method is based on a number of assumptions which affect its accuracy:

Non-dynamic model: A Steady State analysis is for a particular set of conditions. Usually extreme conditions are used, which may not reflect the normal performance of the wall. The Steady State Method does not reflect the effects of time, such as changing temperature and humidity. It assumes that moisture and thermal equilibrium exist. Thermal equilibrium, in some wall systems, can be reached in a few hours. Moisture equilibrium can take months to occur.

__One part of a wall:__ The assembly for which the calculation is performed is the only part of the wall considered. The water vapor characteristics of other wall areas are not considered.

__Direct moisture paths:__ The Steady State Method assumes that all moisture traveling through the wall does so by diffusion. Moisture transfer by air infiltration, such as by cracks, is not considered. This type of vapor flow can be a major source of moisture in walls.

__Multiple incidences of condensation:__ The Steady State Method breaks down after the first instance of condensation because the vapor and thermal properties of the material may be changed by the condensation. Thus the vapor and temperature gradient is affected by the presence of condensation, and the analysis becomes invalid.

__Construction Moisture:__ The Steady State method does not take into account the moisture present due to the construction process.

__Homogeneous Materials:__ The Steady State Method assumes that materials have a constant vapor pressure gradient across their thickness. In fact, very few do.

__Hygroscopy of Materials:__ The ability of a material to take in and give up water is not considered, nor is the change in the vapor or thermal resistance of materials at various water content levels.

Despite the above limitations, the Steady State Method is a common methods of analyzing water vapor flow in construction assemblies. It is an acceptable primary analysis method, especially when coupled with a more detailed analysis if vapor problems appear to exist.

WATER VAPOR PROBLEMS WITH EIFS'S

The composition and materials used in EIFS's make them sensitive to water vapor problems. In particular, the seamlessness of the outside surface tends to reduce the safety margin present in most other types of wall assemblies. Most other assemblies have many small openings in the outside face that allow the vapor pressure to relieve itself to the outside.

In a nutshell, EIFS's must pass water vapor primarily by diffusion. This, coupled with the inefficiency of most vapor retarders that are located on the inside of the wall, can lead to condensation within the wall assembly.

With EIFS's this condensation can occur within the EIFS itself. Since EIFS's are barrier type systems, the condensate has no where to go, and can become trapped within the EIFS.

Similarly, the light construction of the stud and sheathing Substrate System makes it vulnerable to deterioration if high levels of moisture occur. This is in sharp contrast to the European wall structures that are made of unit masonry or concrete. Such Substrate Systems are virtually complete wall systems on their own and can withstand the elements for long periods without significant deterioration.

Thus, it is important that water vapor problems in EIFS's be identified and corrected. If left unresolved, water vapor problems in EIFS's can be very difficult to solve. This is because EIFS wall assemblies do not lend themselves to ventilation or other techniques to equalize the vapor pressure.

The following pages list a number of EIFS wall design conditions that are prone to condensation problems.

CONDENSATION AT LAMINA

Description

Condensation may occur in the vicinity of the Lamina if the vapor resistance of the Lamina is too high. Well-designed EIFS's take this fact into account; the Lamina is specifically engineered to have low vapor resistance. However, sometimes field conditions cause higher-than-intended vapor resistances to actually exist.

Because the Lamina is adhesively-bonded to the insulation, and because the Lamina is cold, the condensate can freeze and expand, resulting in problems with the Lamina.

Figure 2 is in example of condensation at the Lamina due to a Lamina with high vapor resistance. Note the vapor pressures at Interface FG, which is between the EPS and the Lamina. Note also the temperature, which is below freezing.

Solutions

To avoid this, projects using EIFS's should be checked to see if this problem occurs. This should be done in the design stage as it can effect the design of the entire wall assembly. In particular, the use of EIFS Lamina with high vapor resistances should be checked for projects in which vapor flow is normally a problem, such as:

- Extreme humidity on the inside.
- Extremely cold outdoor temperatures.

Vapor retarders can be added on the inside of the wall to reduce the actual vapor pressure near the Lamina.

```
RE: ASTM STP 1107              INSIDE TEMP:    68  | RT TOTAL:    14.060
    Condensation near Lamina   INSIDE RH:      50  | RV TOTAL:     2.034
    ---                        OUTSIDE TEMP:    0  | DIST TO X:    1.37
    ---                        OUTSIDE RH:     30  | TEMP AT X:   26.0
BY: CMD Associates             FILE:    FIG2.WVT   | RATE AT *:    0.03
                                                   |
L   MC   THK    DESCRIPTION                 RT      RV    | LL TEMP VPSAT  VPACT
-   ---  ------ --------------------------- ------ ------ | -- ---- ------ ------
A   41   0.100  Inside Air Film Non-Ref Horiz 0.680 0.010 | AB  65  0.6159 0.3435
B  103   0.050  Latex Paint, 2 Coats        0.000  0.500  | BC  65  0.6159 0.2614
C  141   0.500  Drywall                     0.450  0.014  | CD  63  0.5706 0.2591
D   49   3.500  Air Space Non-Ref           0.910  0.030  | DE  58  0.4879 0.2542
E  147   0.500  Ext Grade Gyp Sheathing     0.450  0.020  | EF  56  0.4511 0.2509
F*  29   3.000  EPS, 1.0 pcf               11.400  1.260  | FG*  1  0.0393 0.0441
G    2   0.150  EIFS Std BC + Finish        0.000  0.200  | GH   1  0.0393 0.0113
H   54   0.100  Outside Air Film Winter     0.170  0.000  | HI   0  0.0376 0.0113
I    1   0.000  ---                         0.000  0.000  | IJ   0  0.0376 0.0113
J    1   0.000  ---                         0.000  0.000  | JK   0  0.0376 0.0113
K    1   0.000  ---                         0.000  0.000  | KL   0  0.0376 0.0113
L    1   0.000  ---                         0.000  0.000  | LM   0  0.0376 0.0113
M    1   0.000  ---                         0.000  0.000  | MN   0  0.0376 0.0113
N    1   0.000  ---                         0.000  0.000  | NO   0  0.0376 0.0113
O    1   0.000  ---                         0.000  0.000  |
```

FIGURE 2. - Condensation Near Lamina

REVERSE VAPOR FLOW

Description

The Lamina must have low vapor resistance to function during winter with warm, humid indoor air and cold, dry outdoor air. Some such sites also have hot, humid summers. Thus, vapor flows outward in the winter and inward in the summer. A breathablr Lamina allows vapor into the wall in the summer, where it may condense.

Figures 3 and 4 are examples of a wall that works in the winter but has stud cavity condensation in the summer. This is due to the high vapor resistance of the interior finish. The same interior finish acts as a vapor retarder in the winter, but restricts the vapor flow during the summer.

Solutions

Vapor barriers on both sides: This can be dangerous since moisture can get trapped in the wall.

Ventilate the wall cavity: This allows the vapor pressure to equalize itself from one side of the wall to the other by direct mass transfer of the moisture, regardless of the vapor flow direction.

Engineering: By judicious selection and positioning of materials, a wall that does not have condensation under winter or summer conditions can be designed.

```
RE: ASTM STP 1107              INSIDE TEMP:      68   | RT TOTAL:    14.060
    Reverse Vapor Flow         INSIDE RH:        50   | RV TOTAL:     5.534
    Winter                     OUTSIDE TEMP:      0   | DIST TO X:
    ---                        OUTSIDE RH:       30   | TEMP AT X:
BY: CMD Associates             FILE:      FIG3.WVT    | RATE AT *:
                                                      |
L  MC   THK   DESCRIPTION                  RT     RV  | LL  TEMP VPSAT  VPACT
-  ---  ----- -------------------------- ------ ------|--- ---- ------ ------
A   41  0.100 Inside Air Film Non-Ref Horiz 0.680 0.010 | AB  65  0.6159 0.3445
B  202  0.080 Vinyl Wall Covering         0.000  4.000 | BC  65  0.6159 0.1032
C  141  0.500 Drywall                     0.450  0.014 | CD  63  0.5706 0.1024
D   49  3.500 Air Space Non-Ref           0.910  0.030 | DE  58  0.4879 0.1006
E  147  0.500 Ext Grade Gyp Sheathing     0.450  0.020 | EF  56  0.4511 0.0994
F   29  3.000 EPS, 1.0 pcf               11.400  1.260 | FG   1  0.0393 0.0234
G    2  0.150 EIFS Std BC + Finish        0.000  0.200 | GH   1  0.0393 0.0113
H   54  0.100 Outside Air Film Winter     0.170  0.000 | HI   0  0.0376 0.0113
I    1  0.000 ---                         0.000  0.000 | IJ   0  0.0376 0.0113
J    1  0.000 ---                         0.000  0.000 | JK   0  0.0376 0.0113
K    1  0.000 ---                         0.000  0.000 | KL   0  0.0376 0.0113
```

FIGURE 3. - Reverse Vapor Flow (Winter)

```
RE: ASTM STP 1107              INSIDE TEMP:      68   | RT TOTAL:    14.060
    Reverse Vapor Flow         INSIDE RH:        50   | RV TOTAL:     5.534
    Summer                     OUTSIDE TEMP:     90   | DIST TO X:    1.15
    ---                        OUTSIDE RH:       80   | TEMP AT X:   78.7
BY: CMD Associates             FILE:      FIG4.WVT    | RATE AT *:    0.13
                                                      |
L  MC   THK   DESCRIPTION                  RT     RV  | LL  TEMP VPSAT  VPACT
-  ---  ----- -------------------------- ------ ------|--- ---- ------ ------
A   41  0.100 Inside Air Film Non-Ref Horiz 0.680 0.010 | AB  69  0.7159 0.3465
B  202  0.080 Vinyl Wall Covering         0.000  4.000 | BC  69  0.7159 0.9193
C  141  0.500 Drywall                     0.450  0.014 | CD  70  0.7333 0.9213
D   49  3.500 Air Space Non-Ref           0.910  0.030 | DE  71  0.7698 0.9256
E  147  0.500 Ext Grade Gyp Sheathing     0.450  0.020 | EF* 72  0.7883 0.9285
F*  29  3.000 EPS, 1.0 pcf               11.400  1.260 | FG  90  1.4101 1.1089
G    2  0.150 EIFS Std BC + Finish        0.000  0.200 | GH  90  1.4101 1.1375
H   54  0.100 Outside Air Film Winter     0.170  0.000 | HI  90  1.4219 1.1375
I    1  0.000 ---                         0.000  0.000 | IJ  90  1.4219 1.1375
J    1  0.000 ---                         0.000  0.000 | JK  90  1.4219 1.1375
K    1  0.000 ---                         0.000  0.000 | KL  90  1.4219 1.1375
```

FIGURE 4. - Reverse Vapor Flow (Summer)

FIBROUS INSULATION IN THE STUD CAVITY

Description

Fibrous insulation has high thermal resistance but little vapor resistance. This allows moisture to flow to areas of the wall without the necessary corresponding loss of vapor pressure.

Figures 5 and 6 show two different EIFS walls with the same total R value. One uses EPS only as insulation. The other uses EPS and fibrous insulation. Note the condensation at the outside of the stud cavity in Figure 6. Condensation at this location can wet the fibrous insulation, further reducing its effectiveness.

Solutions

The use of fibrous insulation in lieu of extra EPS to obtain higher R values should be checked on EIFS projects. The use of fibrous insulation may require the use of a vapor retarder on the inside of the wall. If additional insulation must be placed in the cavity, non-fibrous insulation can help reduce this problem.

```
RE: ASTM STP 1107                INSIDE TEMP:    68  | RT TOTAL:   17.860
    Fibrous Insulation in Cavity INSIDE RH:      50  | RV TOTAL:    2.454 @
    Without Fibrous Insulation   OUTSIDE TEMP:   10  | DIST TO X:
    ---                          OUTSIDE RH:     40  | TEMP AT X:
BY: CMD Associates               FILE:     FIG5.WVT  | RATE AT *:
                                                     |
  L  MC   THK    DESCRIPTION                 RT     RV    | LL TEMP VPSAT  VPACT
  -  ---  -----  -----------------------     ----   -----   -- ---- ------ ------
  A  41   0.100  Inside Air Film Non-Ref Horiz 0.680 0.010 | AB  66  0.6395 0.3438
  B  103  0.050  Latex Paint, 2 Coats         0.000  0.500 | BC  66  0.6395 0.2786
  C  141  0.500  Drywall                      0.450  0.014 | CD  64  0.6077 0.2768
  D  49   3.500  Air Space Non-Ref            0.910  0.030 | DE  61  0.5477 0.2729
  E  147  0.500  Ext Grade Gyp Sheathing      0.450  0.020 | EF  60  0.5200 0.2703
  F  29   4.000  EPS, 1.0 pcf                15.200  1.680 | FG  11  0.0646 0.0512
  G  2    0.150  EIFS Std BC + Finish         0.000  0.200 | GH  11  0.0646 0.0251
  H  54   0.100  Outside Air Film Winter      0.170  0.000 | HI  10  0.0629 0.0251
  I  1    0.000  ---                          0.000  0.000 | IJ  10  0.0629 0.0251
```

FIGURE 5. - Fibrous Insulation in Cavity (w/o insulation)

```
RE: ASTM STP 1107                INSIDE TEMP:    68  | RT TOTAL:   17.500
    Fibrous Insulation in Cavity INSIDE RH:      50  | RV TOTAL:    1.299 @
    With Fibrous Insulation      OUTSIDE TEMP:   10  | DIST TO X:   0.72
    ---                          OUTSIDE RH:     40  | TEMP AT X:  35.3
BY: CMD Associates               FILE:     FIG6.WVT  | RATE AT *:   0.19
                                                     |
  L  MC   THK    DESCRIPTION                 RT     RV    | LL  TEMP VPSAT  VPACT
  -  ---  -----  -----------------------     ----   -----   --- ---- ------ ------
  A  41   0.100  Inside Air Film Non-Ref Horiz 0.680 0.010 | AB   66  0.6385 0.3426
  B  103  0.050  Latex Paint, 2 Coats         0.000  0.500 | BC   66  0.6385 0.2195
  C  141  0.500  Drywall                      0.450  0.014 | CD   64  0.6061 0.2160
  D*87   3.500  Glass Fiber w/o Facing       11.000  0.030 | DE*  28  0.1486 0.2086
  E  147  0.500  Ext Grade Gyp Sheathing      0.450  0.020 | EF   26  0.1386 0.2037
  F  29   1.250  EPS, 1.0 pcf                 4.750  0.525 | FG   11  0.0647 0.0744
  G  2    0.150  EIFS Std BC + Finish         0.000  0.200 | GH   11  0.0647 0.0251
  H  54   0.100  Outside Air Film Winter      0.170  0.000 | HI   10  0.0629 0.0251
  I  1    0.000  ---                          0.000  0.000 | IJ   10  0.0629 0.0251
```

FIGURE 6. - Fibrous Insulation in Cavity (w/ insulation)

MAINTENANCE COATINGS

Description

In is common for coatings to be placed on top of an existing EIFS Lamina. This is done for a number of reasons including:

- Changing the color.
- Repairing damaged areas (so the whole wall looks the same)
- Fixing cracks

Care must be taken to insure that the water vapor breathability of the wall is maintained. Certain types of coatings are inherently vapor resistant, and tend to cause condensation to occur near the Lamina. If the vapor resistance is high enough, the amount of water that can be deposited can be large. Figures 7 shows an EIFS that performs satisfactorily as-is. Figure 8 shows the same wall with a highly vapor resistant maintenance coating added to the Lamina. Note the condensation within the EPS layer.

Solutions

The key is to be sure that the proposed retrofit coating will work on the specific building in question. This requires a knowledge of what is already there, the properties of the new coatings, and an engineering analysis. When in doubt, use a coating that is inherently breathable. Many EIFS manufacturers make special coatings for this purpose.

```
RE: ASTM STP 1107            INSIDE TEMP:    68  | RT TOTAL:   17.860
    Maintenance Coating      INSIDE RH:      50  | RV TOTAL:    2.454
    Without Coating          OUTSIDE TEMP:   10  | DIST TO X:
    ---                      OUTSIDE RH:     40  | TEMP AT X:
BY: CMD Associates           FILE:    FIG7.WVT   | RATE AT *:
                                                 |
L  MC  THK    DESCRIPTION                 RT      RV     | LL TEMP VPSAT  VPACT
-  ---  -----  -----------------------------  ------  ------  | -- ---- ------ ------
A  41   0.100  Inside Air Film Non-Ref Horiz  0.680   0.010  | AB  66  0.6395 0.3438
B  103  0.050  Latex Paint, 2 Coats           0.000   0.500  | BC  66  0.6395 0.2786
C  141  0.500  Drywall                        0.450   0.014  | CD  64  0.6077 0.2768
D  49   3.500  Air Space Non-Ref              0.910   0.030  | DE  61  0.5477 0.2729
E  147  0.500  Ext Grade Gyp Sheathing        0.450   0.020  | EF  60  0.5200 0.2703
F  29   4.000  EPS, 1.0 pcf                  15.200   1.680  | FG  11  0.0646 0.0512
G  2    0.150  EIFS Std BC + Finish           0.000   0.200  | GH  11  0.0646 0.0251
H  54   0.100  Outside Air Film Winter        0.170   0.000  | HI  10  0.0629 0.0251
I  1    0.000  ---                            0.000   0.000  | IJ  10  0.0629 0.0251
J  1    0.000  ---                            0.000   0.000  | JK  10  0.0629 0.0251
K  1    0.000  ---                            0.000   0.000  | KL  10  0.0629 0.0251
```

FIGURE 7. - Maintenance Coating (w/o coating)

```
RE: ASTM STP 1107            INSIDE TEMP:    68  | RT TOTAL:   17.860
    Maintenance Coating      INSIDE RH:      50  | RV TOTAL:    7.454
    With Coating             OUTSIDE TEMP:   10  | DIST TO X:   2.70
    ---                      OUTSIDE RH:     40  | TEMP AT X:  43.9
BY: CMD Associates           FILE:    FIG8.WVT   | RATE AT *:   0.12
                                                 |
L  MC  THK    DESCRIPTION                 RT      RV     | LL TEMP VPSAT  VPACT
-  ---  -----  -----------------------------  ------  ------  | -- ---- ------ ------
A  41   0.100  Inside Air Film Non-Ref Horiz  0.680   0.010  | AB  66  0.6395 0.3447
B  103  0.050  Latex Paint, 2 Coats           0.000   0.500  | BC  66  0.6395 0.3232
C  141  0.500  Drywall                        0.450   0.014  | CD  64  0.6077 0.3226
D  49   3.500  Air Space Non-Ref              0.910   0.030  | DE  61  0.5477 0.3213
E  147  0.500  Ext Grade Gyp Sheathing        0.450   0.020  | EF  60  0.5200 0.3205
F* 29  4.000  EPS, 1.0 pcf                  15.200   1.680  | FG* 11  0.0646 0.2483
G  2    0.150  EIFS Std BC + Finish           0.000   0.200  | GH  11  0.0646 0.2398
H  2    0.080  Maintenance Coating            0.000   5.000  | HI  11  0.0646 0.0251
I  54   0.100  Outside Air Film Winter        0.170   0.000  | IJ  10  0.0629 0.0251
J  1    0.000  ---                            0.000   0.000  | JK  10  0.0629 0.0251
K  1    0.000  ---                            0.000   0.000  | KL  10  0.0629 0.0251
```

FIGURE 8. - Maintenance Coating (w/ coating)

SEALANT JOINTS

Description

Figure 9 shows a typical EIFS sealant joint. Note that the back side of the sealant system is open to the stud cavity. The back side of the sealant is at approximately the same temperature as the outdoor air. Thus, condensation is possible since the moisture laden air in the stud cavity may come in contact with the sealant system.

EIFS's use water-based polymer chemistry in the coatings. Typical EIFS sealant joints have the sealant bonded to the EIFS Lamina. If the Lamina gets wet, such as from condensation, the coatings can soften. When softened, they lose strength. Under those conditions the sealant may pull off of the EIFS, allowing water entry into the wall system.

Solutions

Several EIFS manufacturers have special treatments available for the joints to reduce this problem. Another approach is to insulate the EIFS joint from the back side with a material that does not let moisture through it.

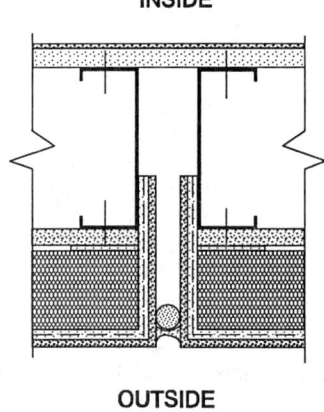

FIGURE 9. - Typical EIFS Sealant Joint

RECOMMENDATIONS

There are a number of things that can be done to improve the performance of EIFS's regarding water vapor, as follows:

EIFS's are a relatively new type of product in the construction market. Information on designing with EIFS's is scarce, particularly about water vapor behavior. To insure that the design community is aware of the water vapor problems presented in this paper, this information should be disseminated. This can be done by articles and papers in trade and professional publications, and by the EIFS manufacturer's product literature.

Similarly, projects that use EIFS's under conditions like those presented in this paper should be checked for water vapor problems as a matter of course. This involves establishing performance standards for the design of EIFS's. Performance standards for EIFS's are being considered now as part of ASTM's EIFS activity.

There currently are no American standards for the performance of EIFS's. The work underway in ASTM to develop such standards should be continued, including standards for water vapor transmission properties of EIFS materials.

REFERENCES

[1] ASHRAE Handbook, 1988 Fundamentals, American Society of Heating, Refrigerating and Air Conditioning Engineers, Atlanta, 1988.

[2] WVT, Water Vapor Transmission Analysis Software, CMD Associates, Vashon, WA, 1988.

Warren R. French, P.E., CCS, CRC

THE PRACTICAL USE AND POTENTIAL LIMITATIONS OF EXTERIOR INSULATION AND FINISH SYSTEM MATERIALS AS AN EXTERIOR BUILDING ENVELOPE

REFERENCE: French, W. R., "The Practical Use and Potential Limitations of Exterior Insulation and Finish Systems as an Exterior Building Envelope," <u>Water in Exterior Building Walls: Problems and Solutions, ASTM STP 1107</u>, Thomas A. Schwartz, Ed., American Society for Testing and Materials, Philadelphia, 1991.

ABSTRACT: The increased utilization of Exterior Insulation and Finish Systems (EIFS) as the primary component of exterior building envelopes for projects ranging in size from small commercial buildings to major high-rise structures has continued unabated for several years now. As new market regions within the United States and the installation work force in these areas are introduced to the use of this building material, there is a natural "learning curve" pertaining to its proper use, limitations, and potential disadvantages. Architects, designers, project supervisors, and local craftsmen are required to acquire a fundamental knowledge of the correct use of EIFS in order to avoid potential problems. This paper presents some of the practical problems being experienced today in regard to this new technology. Although it does not present a specific case study, this paper draws from the experiences of investigation and analysis on dozens of different projects nation wide, some of which were experiencing severe problems related to moisture infiltration. It discusses common mistakes and typical abuses observed in a large percentage of the projects reviewed. In addition, an attempt is made to address correct usage of this increasingly important material by applying well established principles of basic building technology. This paper is intended to assist in achieving the best performance and service life available from this particular building material by disseminating knowledge gleaned from experience with past problems in existing buildings.

KEY WORDS: Exterior Insulation and Finish Systems, EIFS, expansion, contraction, adhered systems

Mr. French is President of French & Associates, a consulting engineering firm located at 15531 Kuykendahl, Suite 275, Houston, Texas 77090.

BACKGROUND

The increased use of EIFS as the primary component of exterior building envelopes for projects ranging in size from small commercial buildings to major high-rise structures has continued unabated for several years now. The reasons for this increase are many, but include the natural implementation of advanced building technology, the search for more economical construction materials, the desirability of wall-cladding systems having increased thermal efficiency, as well as the aesthetic considerations and demands of modern architecture. Another, more simple, reason is the fact that EIFS buildings are visually attractive. EIFS materials provide a plethora of attractive colors and textures, as well as a variety of unique applications, including prefabrication of panels and laser cut foam insulation substrates which allow intricate building ornamentation if desired.

However, as new market regions within the United States, as well as the installation work force in these areas, are introduced to the use of this building material, there is a natural "learning curve" pertaining to its proper use, limitations, and potential disadvantages. As with any new building material, architects, designers, project supervisors, and local craftsmen are required to acquire a fundamental knowledge in the correct use of EIFS in order to avoid potential problems.

Furthermore, with any relatively new technology, it is easy to get carried away with overly optimistic expectations of its use and benefits. Throughout the history of mankind, this optimism has been prevalent in virtually every area of human endeavor, including government, medicine, civil projects, and general industry. To illustrate this point with regard to building construction, it is only necessary to look at the problems encountered during the introduction of single-ply roofing to the United States in the mid-1970's [1],[2]. There were a number of spectacular roofing failures, a few material faux pas, as well as numerous dissatisfied building owners, during the "shaking out" period of single-ply use. Fortunately, despite these setbacks, the single-ply roofing industry has rebounded as manufacturers improved their products, contractors began to understand the differences between various new materials and attempted to educate their labor force, and industry standards were developed for these roofing materials and systems.

Upon review, there appears to be a number of similar parallels between the single-ply roofing industry of a decade or more ago and wall cladding construction of the last several years. Based upon observations of numerous recent projects, it appears that the same angst experienced by single-ply roofing manufacturers and contractors is being repeated by those pioneering the use of EIFS. Unfortunately, if the pattern is completely replicated, a resolution of the problems occurring in relation to EIFS usage will most likely take a similar amount of time and an equivalent number of failed projects before standards are developed for widespread industry use and the construction labor force is sufficiently educated and experienced in the installation of this material.

All of this is to highlight the potential for mistakes and abuses in the utilization of EIFS wall assemblies and (hopefully) disseminate information which will assist in alleviating these problems in the future, as well as to help achieve the performance from this product that should be derived. In general categories, it can be said that the primary mistakes observed with respect to EIFS usage are in regard to design and installation, while the primary abuses are in regard to construction and maintenance. Each of these categories will be more fully developed in the following sections.

DESIGN ISSUES

Building Programming

Obviously, the first series of errors that could occur with regard to the use of EIFS on any project are those related to basic design and architectural programming of the facility. In addition, there are intrinsic material properties that make some products more suitable than others for use in certain applications. Examples are numerous; for instance, storage structures for certain dry chemicals are often constructed using treated wood framing in lieu of steel or concrete which would be subject to accelerated corrosion from the caustic materials which they are designed to house. I suspect that a number of steel-frame chemical storage buildings of this type may have been erected before it was concluded that this application was inappropriate. Likewise, there are certain applications which are unsuited for typical EIFS wall cladding. Such applications, depending upon conditions, could include structures subject to high interior humidities, certain applications of air conditioned buildings in humid climates, applications subject to ignition by radiant heat, and any building that could incur physical damage due either to pedestrian access and vandalism or equipment impact. This is not to say that such applications of EIFS are wrong in all aspects. Certainly, if the potential limitations are recognized and accommodated within the design, it may be possible to achieve acceptable performance. By "accommodating" these factors into the design, it is meant that true resolutions of possible conflicts are developed, not just glossed over or ignored.

Water Vapor Diffusion

One example of a fairly significant design limitation for the use of EIFS is in regard to an air conditioned building located in a humid climate, as established by ASHRAE criteria [3]. Based upon a survey of manufacturers' literature, water vapor permeability of the proprietary EIFS finish coats range between 3.2 and 21.8 perms. These reported test results confirm personal observations that indicate EIFS materials are fairly permeable to water vapor. This fact in itself is not detrimental provided that performance of the EIFS assembly is clearly understood by the designer and is not incorporated into a wall assembly having conflicting components. By conflicting components, it is meant that an air-conditioned building in a hot, humid climate should not be designed with components of the wall assembly (including interior finish materials, such as wall coverings) that have a greater vapor resistance (i.e. lower permeance) than the exterior wall layers. This proscription is in keeping with the well-established principle of having the most vapor resistant materials (lower relative permeability) <u>toward</u> the vapor source, which in this case consists of the exterior, hot, humid climatic conditions. The design significance of this principle would typically indicate that, for air-conditioned buildings in humid climates, low permeance vinyl wall coverings should <u>not</u> be used in conjunction with standard EIFS assemblies that have relatively high permeance to water vapor.

This is a notable contrast to buildings located in cold climates, in which it would be desirable to have a high permeance interior finish material in conjunction with a "breathable" exterior skin. In retrospect, it is not surprising that EIFS type materials were developed and utilized for some time in Europe (with its generally cooler climate) without negative experiences. The experiences of EIFS usage along the Gulf Coast of the southern United States, however, is another matter entirely.

In all design cases, it is obligatory for the designer to determine the need for and optimum position of vapor retarders. In addition, the designer must select appropriate materials, specify, and detail those assemblies and their interface with intersecting walls, ceilings, floors, and other interruptions. In the past, manufacturers have avoided depicting such components within their technical literature because they did not want to take responsibility for them [4]. Now, however, several manufacturers of EIFS materials have begun to offer design consulting services for their products via long distance WATS lines.

Construction Joints

With regard to the practical use of EIFS, it is necessary to deal with another subject which also has potentially controversial aspects. The subject referred to pertains to the suggestion by product manufacturers that construction joints need only be incorporated into EIFS wall surfaces at specific locations of substrate change and changes in the building configuration, and not within expanses of the wall for the purpose of controlling thermal expansion and contraction. Virtually all of the manufacturers' literature reviewed in relation to EIFS construction purports the utilization of EIFS without construction joints at regular intervals, as would be required with more conventional construction materials, such as masonry, precast, or true stucco. Indeed, this design feature is touted as a benefit of using EIFS for modern building design. I suspect that this recommendation by the manufacturers is based upon the flexibility and elongation characteristics exhibited by the proprietary finish system materials, which can be readily tested and evaluated, as well as the European experience of EIFS usage on projects that were predominantly constructed over more rigid, masonry substrates [4],[5].

The American adaption of EIFS usage has been developed using a lighter weight overall construction, usually consisting of a gypsum board substrate over metal stud framing. In addition, as the use of EIFS has expanded and grown, it is common today to find EIFS wall cladding specified for taller, more flexible buildings. The flexibility of these structures could be the result of one or more factors, including the use of computers to derive more efficient structural framing systems, weight saving fabrication and erection procedures, as well as the elimination of heavy weight shear walls, as would be encountered with more traditional concrete and masonry construction. Based upon the common manufacturer's recommendation that construction joints are not needed, there have been a number of mid-rise buildings designed with a continuous vertical expanse of EIFS, extending from the ground floor to the roof parapet. I have personally observed continuous EIFS walls over 61 meters (200 ft.) tall. My experience with EIFS has shown that the elimination of construction joints within EIFS assemblies is an example of an overly optimistic expectation for its practical use. This opinion is based upon observations of cracking within EIFS installations on a number of buildings (see Fig. 1 and 2). Although cracking within EIFS materials my be caused by any number of problems, I have observed that EIFS panels of significant size are more susceptible to cracking and generally less forgiving than may be inferred from the manufacturer's literature. In addition, consideration should be given to the following factors.

First of all, the adhesive/base coat of the proprietary EIFS finish is predominately cementitious, consisting of polymer-modified portland cement. Although the overall material properties are greatly affected by the presence of reinforcing mesh in the composite, if nothing else the EIFS material would require appropriate provision for thermal expansion and contraction in accordance with its particular expansion coefficient. The anticipated expansion referred to here is

FIGURE 1

FIGURE 2

FIGURE 3

FIGURE 4

independent of the material thickness and would occur along wall planes extending throughout the building height and width. The same observation would be true regarding the foam insulation, gypsum sheathing and metal studs, although temperature extremes experienced by the wall components located inboard from the insulation layer would be somewhat moderated. Also, to eliminate all construction joints, it would be required that the EPS foam insulation be dimensionally stable, and not be subject to shrinkage due to heat aging.

Secondly, as construction is completed and the building occupied, increasing dead loads and floor live loads will be imposed upon the building frame. The structural response to these loads will most likely result in elastic column shortening, as well as material creep (if the structural frame is concrete), with the overall consequence of slight changes in the floor-to-floor building height. Although individual floor height changes may not be of great magnitude, the cumulative effect over a long expanse of EIFS (such as in a tall building) could be significant, particularly when considered in conjunction with potential thermal movement.

Finally, with respect to mid-rise EIFS buildings, the prospect of lateral drift occurring at upper floors of the structure in response to wind and seismic design forces is well known. The lateral drift of very tall office buildings under design wind pressures can be several feet. This author was involved in evaluating cladding of an eleven story, concrete framed building located in a hurricane prone coastal area in which the upper story lateral drift under design wind loads was calculated to be approximately 10.2 cm (4 inches). The repetitive compression and tension forces occurring within a continuous (non-jointed) EIFS wall cladding on either side of the building's neutral axis could be significant during such structural dynamics.

The ability of the entire EIFS installation (not just the flexible finish coat) to accommodate such movement and stresses in response to the various conditions described above without utilizing construction joints similar to those used in more traditional wall cladding is highly questionable. My experience indicates that such construction joints should also be used with EIFS designs, then filled with an elastomeric joint material.

Leakage Control Theory

It is possible that one of the factors included in the manufacturer's recommendation to eliminate construction joints is the desire to achieve a "seamless" wall construction. The desire for a seamless wall construction is consistent with the fundamental design concept of EIFS wall cladding, as well as some other modern wall cladding designs which conceptualize a "barrier wall" or single line of defense against water penetration. By a "single line of defense", it is meant that secondary waterproofing, through-wall flashings at each floor, other leakage collection devices, or drainage weeps are not typically incorporated into the wall system. This type of design requires fabrication and assembly of the entire building envelope in a manner which is 100% effective against water penetration (once again, a parallel to single-ply roof construction). My experience indicates that such design concepts for exterior wall systems may be too optimistic, and in fact, impractical for modern construction in which long term performance is required. This opinion has been developed in accordance with previous experience on, not only EIFS walls, but also other wall systems utilizing a "barrier wall" concept, including steel-supported thin veneer stone and precast concrete.

First of all, there will be no building ever constructed for human occupancy which is truly "seamless." At the least, there will always be openings required for means of access. In addition, ventilation, plumbing, and power requirements will also necessitate mechanical openings and interface with exterior spaces for air intakes and exhausts, heat exchangers, etc. Designs incorporating natural light will include windows and skylights, and roof systems must be properly constructed and detailed with respect to vertical elements of the building envelope (see Fig. 3). Furthermore, workmen cannot typically start and stop the application of EIFS finishes in the middle of a panel and maintain appropriate aesthetic effect. Since this is so, it would be prudent for the designer to plan for such interfaces in order to terminate and tie in the production of a subsequent day's work. For all of the reasons enumerated above, it would be wise for the building designer to carefully detail these numerous interface and transition boundaries with other materials.

Window Flashings

Windows particularly need additional design care, since EIFS manufacturers have not traditionally depicted flashing designs within their technical literature [4]. This was another area that was left up to the building designer, however this omission from the manufacturers' typical details led some designers to infer that window flashings were not needed at all. Despite decades of good construction practice with regard to other types of curtain wall materials, as well as a proven need to collect and weep water leakage around window openings using head and sill flashings, a number of buildings have been designed and constructed without such devices.

It would appear that trained professionals more than once exhibited a tendency for overly optimistic expectations in regard to this new technology and, in the process, overturned the wisdom of well-established design criteria and construction practices. In such cases, once construction is completed and leak problems begin to appear, it is virtually impossible to provide acceptable, long-term alternatives short of removal and replacement. Furthermore, if leakage has occurred within the wall assembly, it would be difficult to assess to what extent and how severely ensuing water damage has affected structural integrity of the EIFS assembly.

INSTALLATION ISSUES

Workmanship

One of the most consistent, recurring problems observed on the projects that have been inspected is the lack of proper workmanship achieved by those installing EIFS wall assemblies (see Fig. 4). The workmanship problems range from overt disregard for installation requirements of the manufacturer to simply not planning and coordinating the work with other trades. A number of projects have had terminations and penetrations that were installed contrary to the manufacturer's requirements (see Fig. 5 and 6). Other interruptions within the building envelope were just not detailed by the designer and subsequently were installed by workmen required by necessity to be innovative without proper training and experience to guide them.

Based upon practical field experience, it is mandatory that workmen installing EIFS be knowledgeable regarding standard details for panel terminations at door and window openings, at ground floor slabs, at roof and sheet metal interfaces, and at common penetrations, such as piping, conduit, handrails, etc. (see Fig. 7, 8 and 9).

FIGURE 5

FIGURE 6

FIGURE 7

FIGURE 8

Construction Sequencing

Some installations have been observed in which sequencing of the EIFS installation with other major building components was either not planned or poorly planned regarding the work that had to be performed. In many instances, installation of windows prior to EIFS finishes may increase the difficulty of adequately applying finishes at the panel edges near window perimeters, as well as in achieving 100% embedment of the mesh. Alternatively, rather than maintain a proper joint space, untrained workers will bring the adhesive base coat right up to (and in some cases onto and over) adjacent metal components, such as window frames and sheet metal flashings, allowing no provision for differential movement between these dissimilar materials (see Fig. 3).

Such deficiencies highlight the fact that the limitations of practical job site logistics, as well as the talent and skill of local installation crews, should not be overlooked when designing the EIFS wall cladding, particularly if ornamentation or intricate forms are involved. The contractor responsible for general construction should understand the problems and intricacies of EIFS installation and fully coordinate related construction components with regard to the curtain wall. The EIFS contractor should make sure that the general contractor understands what is involved in his work and take the initiative when necessary to protect the interests of the EIFS installation.

CONSTRUCTION ISSUES

Worker Education

Because of the potential for abuse during construction, it is mandatory that workers located in each new market region are thoroughly educated regarding the proper use of EIFS. The construction education process should not just involve demonstrating EIFS application techniques to crews having a direct association with the wall cladding installation. It should also include appropriate exposure and education of interfacing crafts, such as electricians, window installers, waterproofers that are responsible for sealants, sheet metal workers, landscapers and others who have to do work or install appurtenances in conjunction with the EIFS.

Based upon experience, the workmen of other crafts can create a number of problems by not understanding basic EIFS usage and construction, and by violating the concept of "single line of defense" (see Fig. 8 and 9). Often in these cases, retrofit applications and remedial work will not provide the aesthetics or performance longevity of assemblies that could have been otherwise initially constructed with proper training and planning.

Integrity of Surrounding Components

Another abuse arising from construction aspects of EIFS usage is related to the surrounding components utilized in conjunction with EIFS assemblies. For example, it was previously discussed that penetrations through the wall, such as doors, windows, mechanical equipment, etc., must be carefully designed and integrated into the overall building envelope. Furthermore, the edges of EIFS terminations should be carefully finished and appropriate backing materials and sealants installed in order to provide proper waterproofing. In the event that these components are deficient, the likely result will be water leakage into the EIFS wall assembly, if not into the building interior.

Moisture intrusion into EIFS assemblies has been the concern of many parties for some time. Recent evidence indicates that water damage to gypsum sheathing substrates of EIFS installations can result in a reduced resistance to outward acting forces of common design wind pressures. In addition, deterioration of other interior building components, such as metal studs, fasteners, interior gypsum board, wall coverings, and carpet may be common with profuse or prolonged water leakage. For these reasons, damage resulting from moisture intrusion, whether by direct water leakage or by moist air migration, should be adequately addressed during construction. Even during construction, the amount of time that exterior gypsum sheathing is exposed to the weather may have delayed effects on performance of the wall panel.

MAINTENANCE ISSUES

Physical Damage

One of the beneficial characteristics of EIFS construction is the fact that installed materials require little or no maintenance. That is assuming, of course, that the EIFS assembly has been installed properly and that no damage has been incurred since its completion. A number of cases have been investigated in which physical damage to EIFS wall components and finishes have figured prominently. Physical damage can occur during construction as panels are erected or subsequent to construction during building operation and use.

The most common types of physical damage are typically related to poor building use and maintenance procedures. For example, the installation of signs and other appurtenances through the EIFS skin, if not performed properly, can result in water leakage. Also, motorized swing stages utilized for window washing or other maintenance operations can cause damage on the face of building elevations, as well as at unprotected parapets, such as those using a manufacturer's EIFS detail on its top surface (see Fig. 10). A final area to be aware of is in relation to exterior grades that have been built up too high around EIFS walls at ground floors and at planter boxes. Planting, mulching, and cultivating activities are typically carried out as an on-going activity, and usually involve the use of spades, rakes, and hoes that could damage the EIFS finish. To avoid this potential abuse, it will be necessary to instruct landscaping contractors and lawn maintenance crews in the care of EIFS construction. If EIFS materials are intentionally used by the designer as a planter liner, supplementary waterproofing of underlying substrates should be included.

Another situation that can commonly result in damage to installed EIFS assemblies is when subsequent renovation work is performed. Particular care should be used whenever cleaning procedures are used on the EIFS surface. Excessively abrasive cleaning methods, such as sand blasting, should not be used; and water blasting would be ill-advised in most cases. It is recommended that the EIFS manufacturer be consulted for the most appropriate cleaning method for each individual project. In addition, when post-construction sealant renovation is required, it would be prudent to restrict the use of heavy solvent washes to effect cleaning of the sealant joints. At a few projects, excessive solvent-use on the EIFS panel has resulted in vapors that permeated the finish coat (leaving it unharmed) and attacked the underlying expanded polystyrene (EPS) insulation board. At least one project was subject to core cuts through the wall panels, in which it was found that large voids had formed within the EPS, leaving the proprietary EIFS layers completely unsupported. These voids developed as the EPS dissolved in reaction to the solvent vapors.

74   WATER IN EXTERIOR BUILDING WALLS

FIG. 9

FIG. 10

RECOMMENDATIONS

The following recommendations are presented as a summary of items related to, and growing out of, the cumulative inspections and observations related to EIFS usage in the United States over the past several years. Obviously, such a summary cannot be comprehensive, since each architectural project will have its own individual design considerations. However, it is intended that this summary be utilized as a check list of some of the matters to be considered during design and construction of a project which is to include EIFS components.

Related to Design

1. Evaluate the anticipated building occupancy and use for the potential of pedestrian vandalism or vehicular damage. If these factors will be matters of concern, consider the use of true stucco, a polymer-modified (PM) EIFS, or at least PB EIFS with high impact mesh or multiple layers of reinforcement.

2. Check the relative permeabilities of wall materials at each exterior wall section. Be sure that both winter and summer design conditions are considered, and remember to apply good principles of design regarding dew point, condensation, moisture accumulation, and the use of vapor retarders and air barriers. EIFS manufacturers could help in this regard by reporting permeance values in uniform units. It was noted that various product data reviewed in preparation for writing this paper reported water vapor transmission (WVT) performance of these materials by using units of WVT, permeance (some using metric units), and permeability. Only one manufacturer reported the specific method or procedure utilized within the six available from ASTM E96. In addition, several other test methods were used by some manufacturers, the results of which would not be directly comparable to

those obtained using the ASTM E96 test standard. Proposed industry standards currently being developed would be well advised to require a standardization of testing method and reporting of units in order to facilitate comparisons. If the building designer does not feel comfortable performing the proposed evaluation or there are special conditions involved, get outside help which may be available from the EIFS manufacturer, consultants, or technical papers on the subject.

3. Consider the possibility of providing air barriers or otherwise sealing the intersection of interior partitions at exterior walls, particularly for hotel buildings where moist air can travel into these partitions.

4. For multiple story buildings and facilities having a large horizontal plan dimension, use true expansion joints occurring regularly within the surface of the EIFS. I generally recommend that vertical true expansion joints be located approximately 6 meters (20 feet) from building corners, and spaced about every 12 meters (40 feet) thereafter. In addition, horizontal joints should be placed within the cladding of multi-story buildings at regular intervals throughout its entire height. I have had success detailing true expansion joints at every second floor level, however, some authorities recommend them at every floor. Please note that the size of these joints will depend upon many factors, including the coefficient of thermal expansion of the particular EIFS material utilized, as well as physical properties of the sealant used, general building design, etc. Last but not least, all recommendations of the manufacturer regarding use of joints at building ells, tees and U's, as well as at changes within the substrate should be strictly adhered to.

5. Despite the depiction by certain manufacturers of various details, remember that, ultimately, the building designer is responsible for design of the facility which is to use EIFS. Bring past training and experience to bear upon the use of this material, just as you would any new product. Give particular attention to the design of penetrations through the EIFS, including doors, windows and their various flashing requirements. If there is the slightest doubt, involve the manufacturer or other knowledgeable consultant in developing special details. Avoid the use of parapets or knee walls that are constructed with EIFS copings or top surfaces; they are subject to physical abuse and can allow subsequent water entry.

6. Prior to releasing construction documents, survey the building elevations for potential problem areas; try to anticipate openings and penetrations that may not specifically appear on the drawings, such as electrical conduit, lighting fixtures, sign attachments, downspouts, etc. Provide typical details of these penetrations within construction documents and enforce necessary compliance.

7. Consider the use of EIFS wall assemblies that utilize mechanical fastening of the insulation layer. There is no conclusive evidence of an advantage related to solely depending on the bond of the proprietary adhesive, and there is a definite disadvantage to this method of securement in the event that water leakage is experienced to the degree that deterioration of the gypsum sheathing is incurred.

8. Consider the viability of specifying prefabrication of EIFS panels and components whenever practical. The level of workmanship tends to be improved in a controlled environment, and quality control procedures typically can be more readily implemented and consistently maintained in a "factory" setting. Obviously, this alternative will require a somewhat more detailed design with

respect to pre-construction planning, fabrication shop drawings, erection drawings, etc. However, the additional thought, coordination, and planning arising from these activities generally result in a better end product overall. The designer should specify strict curtain wall performance requirements, and prequalify EIFS fabricators in order to make certain that adequate fabrication facilities and experienced personnel are available. Also, it is recommended that there be a single source responsibility for panel fabrication and erection (and possibly even application of sealants) for each pre-fabricated panel project.

Related to Installation

1. Make certain that all personnel involved in the EIFS installation are adequately trained and competent to perform their assigned tasks. Supervisors should be sufficiently experienced with EIFS applications to allow close monitoring of the work.

2. Establish an internal quality assurance program regarding installation procedures and tolerances. Particular attention should be given to monitoring the application of finishes at panel edges, terminations, and penetrations. Backwrapping of the reinforcement fabric and proper edge finishing are material requirements that are also commonly lacking and need special care.

3. Maintain proper clearances between unfinished edges of foam insulation board and adjacent components so when finishes are applied adequate joint tolerances are retained. This recommendation applies to joints occurring panel-to-panel, panel-to-door, panel-to-window, etc. Also, make sure that EIFS finishes are not built-up to excessive thicknesses at aesthetic joints.

4. Coordinate EIFS installation with the work of other trades. Utilize a pre-installation conference with all related trades present so that installation sequencing, procedures, proscriptions, and interface points may be discussed and resolved. Other trades that may be included in this conference would be the metal stud framing erectors, gypsum sheathing installers, roofing contractor, sheet metal contractor, electricians, plumbers, and air conditioning contractors.

5. Closely inspect joint construction and interfaces with adjacent appurtenances. If there are no details, and no design direction included within the project documents, then get assistance from the manufacturer or request clarification from the architect.

6. Explicitly follow all of the manufacturer's installation recommendations. Coordinate and obtain manufacturer's input and approval for any special conditions.

7. Avoid prolonged exposure of unprotected gypsum sheathing to the weather. Temporarily protect the sheathing board from moisture if there is going to be a delay installing EIFS materials.

Related to Construction

1. For all of the building systems, follow through with quality assurance programs which may have been instituted. Resist the temptation to take short cuts which could affect long term performance of the EIFS. Also, conduct a training program for key construction personnel in order that the proposed EIFS installation, its performance, and limitations are at least somewhat familiar to these workers.

2. Make certain that flashing assemblies are correctly constructed and coordinated between the trades of related work. Also, plan and schedule overall construction so that interfacing components are sequenced properly.

3. Sealant installers should make sure that all EIFS surfaces upon which they apply their materials are properly and completely finished. If deficiencies exist, interrupt sealant work until corrections can be made. Make certain that all openings that could allow direct water leakage (and even moist air leakage in humid climates) are thoroughly and appropriately sealed.

4. Protect the EIFS installation from moisture intrusion and physical damage during construction. Control access to scaffolds and stages around the EIFS work and make sure that any damage is reported and properly repaired immediately.

Related to Maintenance

1. Owners and property managers should closely monitor any activities that would affect the performance of installed EIFS assemblies. An inspection of the building exterior should be made before and after performing maintenance work that requires use of a motorized swing stage. If any damage is discovered, immediate repairs should be made by an approved EIFS contractor. Periodically, an inspection should be made of exterior building walls at ground level to ascertain damage from planters and landscaping or from lawn keeping equipment.

2. Establish a program of inspecting and maintaining the exterior building envelope on regular, periodic intervals, for example, every two years (or even more often). These inspections should be performed using equipment that allows close visual inspection, and should include an evaluation of the overall EIFS condition, weathering of and defects within sealants, and condition of adjacent building components, such as roofs, doors, windows, flashings, etc.

3. Whenever maintenance or repairs to the building envelope are required, make sure only qualified personnel are employed that are experienced in working with EIFS. Such work might include roofing repairs or replacement, sealant renovation, window washing, etc.

ACKNOWLEDGMENTS

A special thanks goes to Sharon Laird for review and editing of this manuscript.

REFERENCES

[1] Rossiter, Jr., Walter J., "Single-Ply Roofing - A Decade of Change," <u>ASTM Standardization News</u>, American Society for Testing and Materials, Volume 13, Number 9, September 1985; p. 32-35.

[2] Cullen, William, "Project Pinpoint Finds Single-Plies Dominating: Problems with Newer Materials Rise with Use," <u>Roofing Spec</u>, National Roofing Contractors Association, August 1987; p. 19-20.

[3] "Moisture in Building Construction," in <u>ASHRAE Handbook of Fundamentals</u>, 1985, p. 21.17-21.19.

[4] Russell, James S., "EIF Systems: quality lost in the translation?," <u>Architectural Record</u>, July, 1989; p. 124.

[5] Doyle, Margaret, "Trends in specifying EIFS," <u>Building Design & Construction</u>, Volume 29, Number 8, August, 1988; p. 60.

Warren R. French, P.E., CCS, CRC

IN-SITU TESTING OF THE STRUCTURAL INTEGRITY OF
EXTERIOR INSULATION AND FINISH SYSTEMS

REFERENCE: French, W. R., "In-Situ Testing of the Structural Integrity of Exterior Insulation and Finish Systems," Water in Exterior Building Walls: Problems and Solutions, ASTM STP 1107, Thomas A. Schwartz, Ed., American Society for Testing and Materials, Philadelphia, 1991.

ABSTRACT: Utilization of Exterior Insulation and Finish Systems (EIFS) as exterior cladding for modern commercial buildings is increasing. Although adequate test methods exist for determining physical and material properties of individual components that make-up EIFS, there are presently only limited methods for evaluating the performance and structural integrity of EIFS cladding that has been in place for any period of time. This paper discusses the results of informal testing performed on several different types of buildings located in various parts of the country. The testing procedure was destructive and evaluated the resistance of existing assemblies to applied pressures on the vertical wall surface. Varying results were obtained since a number of the wall components had experienced deterioration related to past water leakage. Test results emphasize the need for formal tests of this nature to be developed and highlight the critical nature of protecting certain portions of the EIFS assembly from moisture damage.

KEY WORDS: Exterior Insulation and Finish Systems, EIFS, moisture damage, design wind pressures

BACKGROUND

General Information

The utilization of non-traditional, exterior cladding systems in the construction of modern commercial buildings has experienced a significant increase in the past several years. Some of the reasons for this experimentation and innovation include the search for more economical materials and systems for use in speculative real estate, as well as the implementation of recent advances in building technology. Among the numerous alternative materials enjoying widespread use by architects and contractors today is what has become generically classified as Exterior Insulation and Finish Systems (EIFS).

Mr. French is President of French & Associates, a consulting engineering firm located at 15531 Kuykendahl, Suite 275, Houston, Texas 77090.

Polymer based (PB) EIFS wall panels may be prefabricated or built in place and typically consist of a gypsum sheathing board mechanically fastened to studs, expanded or extruded polystyrene insulation board (either adhered to the sheathing or fastened to the underlying framing), and a proprietary finish system. The proprietary finish system is generally composed of an adhesive/base coat reinforced with a glass fiber fabric, as well as an acrylic-based finish color coat. Although there have been installations in service in the United States for as long as twelve to fifteen years (and even longer in European countries), to date there has been a general lack of field evaluation methods developed for use with these types of wall systems.

Presently, there are recognized industry standards for evaluating components of the EIFS, such as compression strength and water absorption of the polystyrene insulation board, elongation and tensile strength of the acrylic finish coating, etc. There are, however, no standards to evaluate "system" integrity, particularly for installations that have been in place for some time [1].

This paper presents a description of informal tests developed for use on several buildings that had already been in service for a number of years. Three hotel buildings located in various parts of the country and utilizing different construction methods were subjected to the test method. All three structures had experienced varying levels of performance distress, particularly related to water leakage and moisture intrusion. Two of the three buildings had experienced incidents in the past in which portions of the building envelope had been removed or dislodged by high winds. Each of the owners of these buildings desired to have an objective evaluation of the structural condition of remaining EIFS assemblies in place on the structures. A total of thirty-five (35) specimens were evaluated during the course of this test program.

Description of Projects

Midwest State: This project was a five year old, mid-rise, concrete frame structure located in a midwest state. A design wind pressure of 910 Pa (19 PSF) had been determined for this project in accordance with applicable codes. EIFS wall panels for this job had been fabricated on site. Gypsum sheathing had been secured to metal studs using screw fasteners consistently spaced at 30.5 cm (12") on center. The proprietary finish had been applied over 3.8 cm (1½") of expanded polystyrene foam insulation board that had been partially adhered to the gypsum sheathing in general accordance with the manufacturer's requirements.

Some amount of water damage and moisture related deterioration was known to exist due to visual inspections conducted after a small portion of one wall panel was found loose and in danger of falling. It was suspected that other areas of weakened EIFS components would exist since an inspection of the building exterior revealed common anomalies, such as cracks occurring within the aesthetic joints, as well as defective and failed joint sealants. A total of four (4) specimens (numbers 1 through 4) were evaluated at this project.

Mid-Atlantic State: This project was a seven year old, mid-rise, concrete frame structure located in a mid-Atlantic state. A design wind pressure of 1,436 Pa (30 PSF) had been determined for this project in accordance with applicable codes. EIFS wall panels for this job had been pre-fabricated and erected in place using common fabrication and anchorage methods. Gypsum sheathing had been secured to individual metal stud frames using screw fasteners consistently spaced at 30.5 cm (12") on center. The proprietary finish had been applied over 5.1 cm (2") of expanded polystyrene foam insulation board that had been partially adhered to the gypsum sheathing in general accordance with the manufacturer's requirements.

Some amount of water damage and moisture related deterioration was known to exist at this project due to visual inspections conducted after a small portion of one wall panel was removed by high winds during a storm. It was suspected that other areas of weakened EIFS components would exist since an inspection of the building exterior revealed common anomalies, such as cracks occurring within the aesthetic joints, as well as defective and failed joint sealants. A total of four (4) specimens (numbers 5 through 8) were evaluated at this project.

<u>South Central State:</u> This project was a seven year old, low-rise, concrete frame structure located in a South Central state. A design wind pressure of 2,394 Pa (50 PSF) had been determined for this project in accordance with applicable codes. EIFS wall panels for this job had been fabricated on site. Gypsum sheathing had been secured to metal studs using screw fasteners spaced from 16.3 cm to 30.5 cm (6"-12") on center. The proprietary finish had been applied over 2.5 cm (1") of expanded polystyrene foam insulation board that had been partially adhered to the gypsum sheathing.

A number of areas were found in which the assemblies apparently had not been constructed in accordance with the manufacturer's requirements. In addition, moderate-to-severe amounts of water damage and moisture related deterioration was known to exist due to previous investigations conducted at the site. Common anomalies consisted of water leakage from wall penetrations and windows, defective and failed joint sealants, as well as cracks occurring within the EIFS finish. A total of twenty-seven (27) specimens (numbers 9 through 35) were evaluated at this project.

PURPOSE

The purpose of the test method developed and utilized with respect to these buildings was to evaluate the resistance of existing EIFS installations to negative pressures ("suction"), which can commonly arise due to wind forces acting against the exterior building envelope. One of the main criteria affecting the design of this test method was a desire to provide an apparatus that would be readily portable, in order that numerous tests could be conducted as economically as possible.

APPARATUS

Equipment utilized for these tests consisted of a lightweight aluminum frame, which had been specially fabricated for this purpose from square tubular members (see Fig. 1 and 2). Usage of the test frame required it to be mounted against the vertical wall surface while being temporarily supported from a motorized swing stage (see Fig. 3), or else from scaffold frames erected on the ground. The test frame was fitted with oversized contact plates which were designed to bear against the side of the building, distributing frame reaction forces in order to minimize or alleviate compression of the adjacent surrounding EIFS assembly.

The pull device consisted of a manual, geared winch that had been provided with 5 mm (3/16") diameter aircraft cable. A common hanging scale with a range from zero to 1,780 Newtons (400 lbs.) was used in line to measure applied force. The simulation of negative pressure

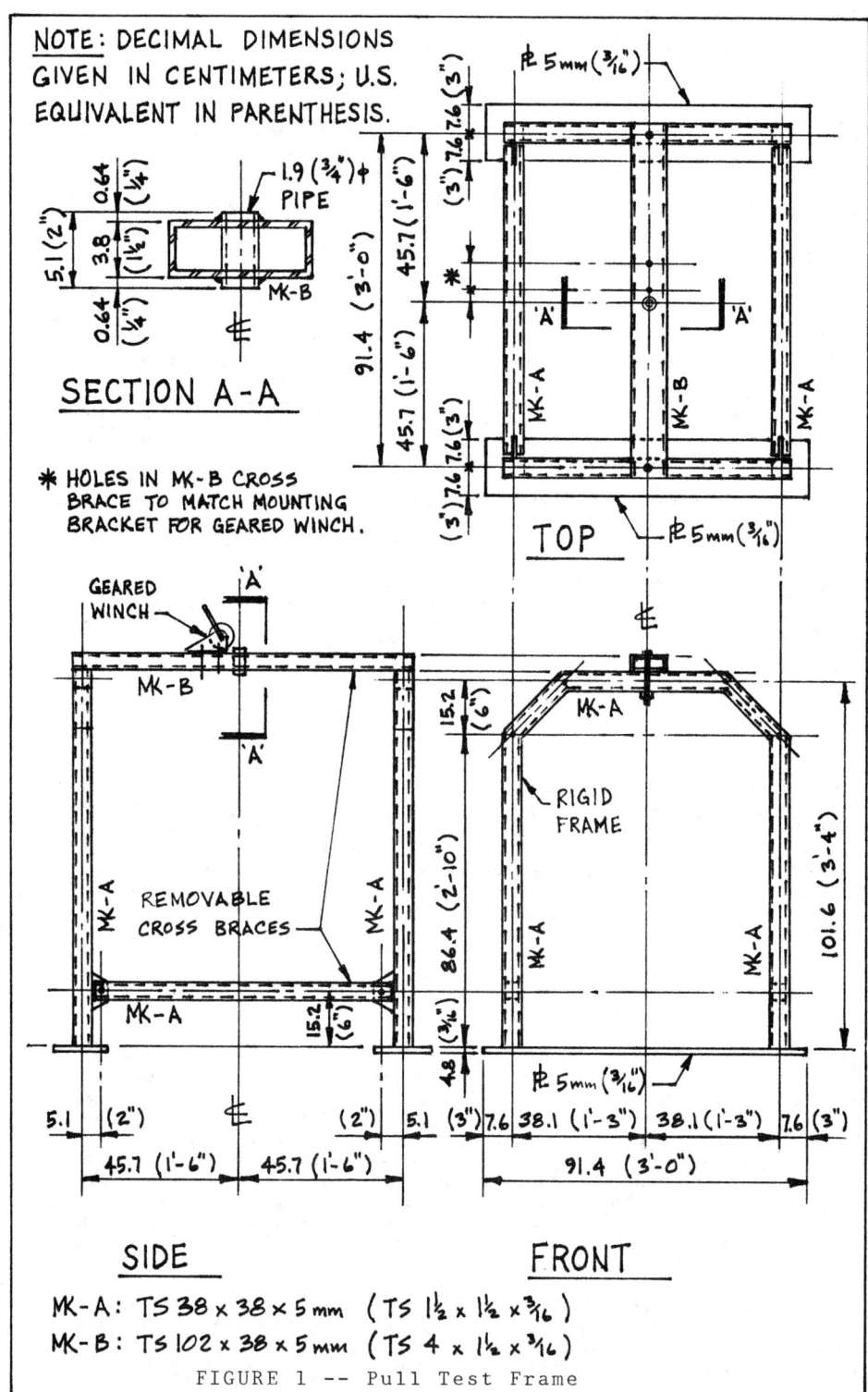

FIGURE 1 -- Pull Test Frame

FRENCH ON IN-SITU TESTING OF INSULATION AND FINISH SYSTEMS 83

FIGURE 2

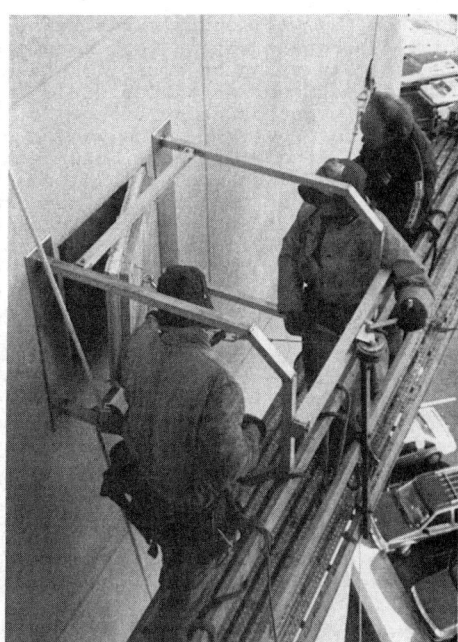

FIGURE 3

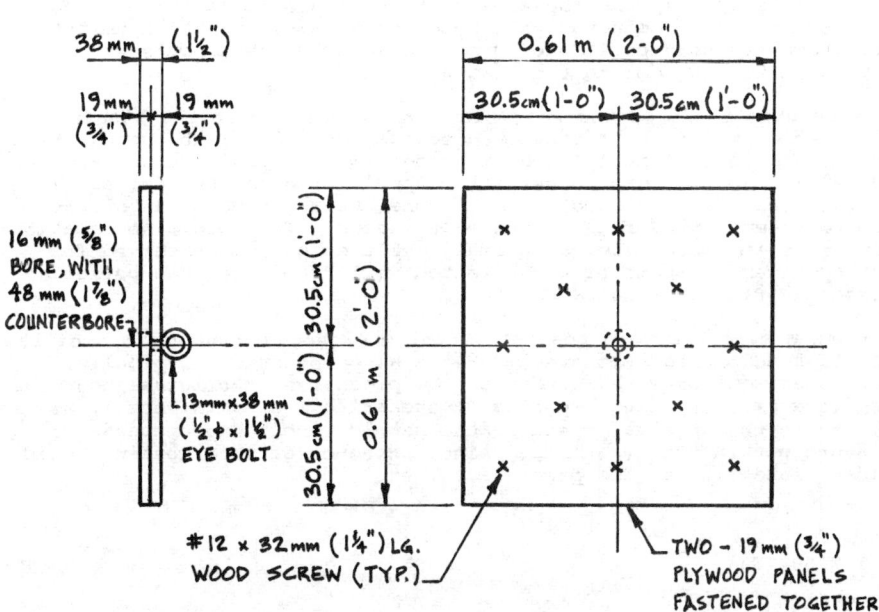

FIGURE 4 -- Wooden Test Module

against the wall surface was transferred to the EIFS assembly via a 61 cm (24") square wooden test module which had been adhered to the EIFS finish coat using an epoxy adhesive. The wooden test modules had been fabricated from two pieces of 19 mm (3/4") thick plywood that had been fastened together using twelve No. 12 x 32 mm (1-1/4") wood screws in a prescribed pattern. The wooden test module was also provided with a 13 mm (1/2") diameter eye bolt secured through the center which was used as the attachment point (see Fig. 4). On a few of the test modules, due to cold outside temperatures, a problem developed with regard to obtaining adequate adhesion using the epoxy. In such cases, the epoxy was allowed extra time to cure, or else four toggle bolts were utilized to mechanically fasten the test module to the EIFS assembly. On the specimens using toggle bolts, the adequacy of cementitious adhesive bonds could not be evaluated, but this was deemed acceptable since determining the condition of gypsum sheathing became more important for these tests. A total of two specimens were tested using toggle bolts in lieu of adhesion only.

It should be noted that the concepts used to develop this test method pertaining to EIFS wall systems drew upon many sources. One of the primary sources was the "pull test" method of performing field uplift tests in regard to built-up roofing systems, which was developed by Factory Mutual Engineering Corporation [2].

PROCEDURE

Prior to conducting the pull test, an individual wooden test module was adhered to a pre-selected exterior wall surface location using an all-purpose epoxy adhesive. The adhesive was allowed to cure an adequate amount of time and temporary support for the wooden module was provided, if necessary, during this process. Exterior building wall panels were saw cut all the way through the EIFS assembly immediately against and around the adhered test module in order to isolate the test area from adjacent wall panel construction. The aluminum test frame was then positioned over the adhered test module with the contact plates straddling the test specimen.

The tension cable of the test frame was connected to the hanging scale, which was in turn connected to the eye bolt of the wooden test module. An initial pull force of predetermined magnitude was applied and held for one minute. Thereafter, pull force on the test specimen was successively increased in one minute intervals using a "stepped" loading scheme until failure was experienced. The load/time relationships were recorded, and the primary failure mode was observed for each specimen. Repair of the affected wall panel area was performed as soon as was practical.

To the greatest extent possible, contemporaneous observations of the overall condition of the specimen were also obtained. Typically, these observations were related to the precision of original construction (e.g. stud spacing, spacing of sheathing fasteners, etc.), as well as to the physical state of the sample, including dampness or moisture within the gypsum sheathing, presence of water stains, mold, mildew, integrity of the paper facer, etc.

TEST RESULTS

The results of these tests varied widely from building to building, as well as between individual specimen locations occurring at the same building. Primarily, these differences could be attributed to the particular history of the test specimen with regard to water leakage and the degree of damage incurred as a result of on-going moisture intrusion processes.

As may be expected, it can be generally stated that samples visually exhibiting the most advanced deterioration (or loss of integrity) due to water damage were also those that exhibited severely deficient performance in regard to resistance against negative wind pressures. Although delamination of paper facers has been of some concern in the past [3],[4], it was noted that the largest number of failures involved screw heads pulling through the gypsum sheathing which was in a weakened condition. This weakening of the gypsum appeared to be due to a loss of structural strength caused by water damage.

TABLE 1 -- Final Failure Loads for Each Specimen

Specimen	Applied Load, Pa	Applied Load (PSF)	Time
1	3,352	70.0	7
2	1,796	37.5	3
3	718	15.0	2
4	3,112	65.0	6
5	2,394	50.0	6
6	4,369	91.3	9
7	4,788	100.0 (max.)	12
8	3,028	63.3	5[a]
9	2,377	49.7	10
10	2,648	55.3	9
11	1,135	23.7	4
12	2,364	49.4	8
13	1,135	23.7	4
14	1,589	33.2	5
15	2,648	55.3	9
16	1,823	38.1	7
17	1,892	39.5	7
18	2,006	41.9	7
19	1,733	36.2	6
20	2,364	49.4	8
21	2,321	48.5	8
22	1,003	21.0	4
23	371	7.8 (min.)	1[a]
24	1,986	41.5	7
25	1,797	37.5	6
26	1,797	37.5	6
27	1,561	32.6	6
28	3,045	63.6	8
29	2,849	59.5	8
30	2,932	61.2	10
31	2,364	49.4	8
32	946	19.8	3
33	2,321	48.5	8
34	1,178	24.6	4
35	1,276	26.7	4

a - failed while going to next load increment

Although the incremental step loadings for each specimen were recorded, for the purposes of this paper it is deemed necessary to only present final loads at failure for each specimen. The mean of the initial applied loads for all samples was 575 Pa (12 PSF), for the first minute of testing, with a standard deviation of 197 Pa (4.1 PSF). The applied loads were then incremented upward at one minute intervals utilizing steps having a mean of 361 Pa (7.55 PSF) and a standard deviation of 58 Pa (1.2 PSF).

Numerical Results

Systems Overall: The maximum pressure successfully resisted by any particular specimen was 4,788 Pa (100 PSF), indicating that the various code approvals claimed by EIFS manufacturers in regard to high wind, coastal areas could be justified for systems installed properly and protected from moisture-related deterioration. However, the lowest pressure recorded was an anemic 373 Pa (7.8 PSF). The arithmetic mean, or median, failure pressure for all thirty-five tests was 2,144 Pa (44.8 PSF), with a standard deviation of 953 Pa (19.9 PSF). Seven of the thirty-five specimens, or 20% of the samples, exhibited failure pressures less than 1,197 Pa (25 PSF). Eleven specimens (31.4%) were between 1,197 Pa (25 PSF) and 2,155 Pa (45 PSF), while seventeen specimens (48.6%) exhibited failure pressures greater than 2,155 Pa (45 PSF).

It is to the credit of these types of systems that a total of seven specimens (20%) exhibited failure pressures greater than 2,873 Pa (60 PSF), which would typically be more than adequate to meet design wind pressures in most commercial structures located in the United States. However, despite the overall performance exhibited by test samples taken as a whole, it was found that a certain percentage of specimens at two of the three projects exhibited results that were deficient when compared to the required design wind pressure. With respect to this criteria, seventeen of the thirty-five EIFS specimens (48.6%) failed to meet the wind load criteria for their geographic settings.

Midwest State: The test results from this project were 3,352 Pa (70 PSF), 1,796 Pa (37.5 PSF), 718 Pa (15 PSF), and 3,112 Pa (65 PSF). The median failure pressure was 2,245 Pa (46.9 PSF), with a standard deviation of 1,226 Pa (25.6 PSF). One of the four test samples (25%) fell below the design wind pressure of 910 Pa (19 PSF).

Mid-Atlantic State: The test results from this project were 2,394 Pa (50 PSF), 4,369 Pa (91.25 PSF), 4,788 Pa (100 PSF), and 3,028 Pa (63.25 PSF). The median failure pressure was 3,645 Pa (76.1 PSF), with a standard deviation of 1,122 Pa (23.4 PSF). There were no test samples that fell below the design wind pressure of 1,436 Pa (30 PSF).

South Central State: The test results from this project ranged from 371 Pa (7.8 PSF) to 3,045 Pa (63.6 PSF). The median failure pressure was 1,906 Pa (39.8 PSF), with a standard deviation of 680 Pa (14.2 PSF). Technically, a total of twenty-two specimens (81.5%) fell below the design wind pressure of 2,394 Pa (50 PSF). It should be noted, however, that six of these were within 3% of the design stress, a variance which is typically considered to be acceptable when using the working stress method of design with an appropriate factor of safety. Accordingly, this project would only be counted as having sixteen specimens (59.3%) which did not meet the specified wind load criteria for this geographic location.

Failure Modes

As each test specimen was taken to failure, observations were made with regard to the failure mode experienced. Since each of the assemblies had experienced some amount of water damage, it was pertinent to record the mode of failure in order to identify those components most susceptible to deterioration. As might be expected, the apparent "weak link" of these systems is the loss of strength occurring within the gypsum sheathing board once it has been subject to progressive water leakage. It is of particular interest to note that as the severity of water damage to the gypsum core increases the more likely it becomes that paper facers utilized at the exterior face of the sheathing contribute to the overall failure mechanism.

For all specimens, it was noted that 57.1% exhibited a failure mode consisting of the attachment screws pulling through the gypsum sheathing. Generally, this failure type (the largest single category) resulted in a conical-shaped shear plane, consistent with models utilized for embedment analysis in concrete. Failure attributed to facer delamination acting alone represented only 8.6% of the samples included within this test program, while inadequate bond of the cementitious, adhesive/primer occurred only once (see Fig. 5).

The second most prevalent failure type were those that consisted of a combination of fastener pull-through with facer delamination or some other element, such as primer disbonding. These types of failures represented 31.4% of all samples tested. Of the combination type failures, it was observed that some could be attributed more to fastener pull-through (63.6%), which was observed to have occurred over more than half the surface area of each individual specimen. The failure of other samples could be attributed more to facer delamination (18.2%), which was observed to have occurred over half the surface area of those specimens. Overall, these two kinds of combination-type failure modes represent 20.0% and 5.7% of the total test program. It was also noted that one specimen exhibited a failure involving what was estimated to be a 50-50 split between the pull-through and facer delamination modes, and one other specimen exhibited a failure involving pull-through of a portion of the sample, with disbonding of the adhesive/base coat on the remainder of the sample surface.

Geographically, the results were not that revealing except that all of the combination-type failure modes occurred at the South Central state. This condition could simply be related to the construction materials and methods used at this project, or it may be indicative of more complex problems related to the water leakage or high ambient relative humidities experienced at this particular location. In addition, studies of this project, which could be classified as an air conditioned building in a humid climate in accordance with ASHRAE criteria [5], revealed that summer season condensation, as well as accumulation of moisture within exterior walls, was wide spread and prevalent. For all projects, it was also noted that the orientation of various building elevations with respect to compass bearings (i.e. north, south, etc.) did not appear to have any effect on the number of failures or the structural integrity displayed.

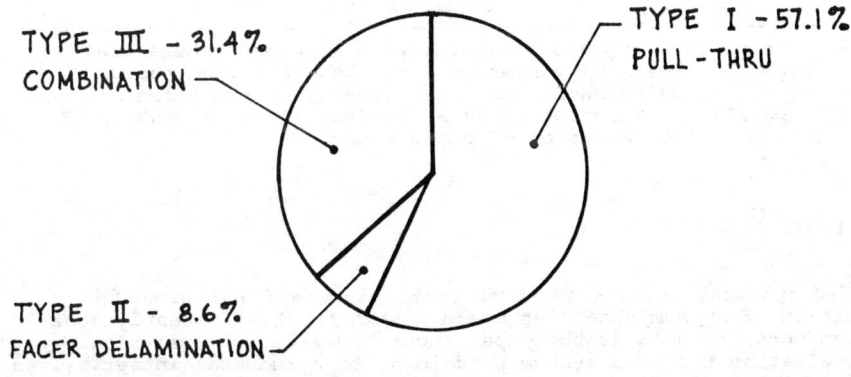

FIGURE 5 -- Distrubution of Failure Modes

TABLE 2 -- Failure Modes

Specimen	Failure Mode[a]	Comments
1	Type I	No evidence of leakage observed
2	Type I	Previous leakage, mold
3	Type I	Previous leakage, mold, sheathing damp
4	Type I	No evidence of leakage observed
5	Type I	Severe water damage, sheathing damp
6	Type I	Random stains, slight corrosion
7	Type I	Slight corrosion
8	Type I	Water damage, some corrosion
9	Type I	No evidence of leakage observed
10	Type I	Water stains, some mold
11	Type III	Water damage, mold, corrosion
12	Type III	Dry, no corrosion
13	Type III	Water damage, mold, corrosion
14	Type III	Dry, no corrosion; improper adhesive
15	Type I	Dry, no corrosion
16	Type III	Water stains, mold
17	Type I	Sound & dry; partial pull-thru @ 31.6
18	Type I	Slight corrosion
19	Type I	Corrosion
20	Type III	Water stains, mold
21	Type I	Water stains, mold, diagonal crack; partial pull-thru @ 36.9 PSF
22	Type I	Water stains, mold, corrosion
23	Type II	Severe water entry at facer, corrosion
24	Type I	Sound and dry
25	50-50 Types I & II	Water stains, mold, corrosion
26	50% Type I	50% adhesive/base coat disbond
27	Type II	Facer stained and discolored
28	...	Sound & dry, adhesive/base coat disbonded
29	Type I	Water stains, mold
30	Type I	Water stains, minor corrosion
31	Type I	Minor corrosion
32	Type III	Minor corrosion
33	Type III	Sheathing stained, minor corrosion
34	Type III	Mold, minor corrosion
35	Type II	Mold, minor corrosion

a - Three primary types of failures were recorded, these are:
    Type I: Heads of screws pulling through gypsum board
    Type II: Delamination of paper facer from sheathing
    Type III: Combination of screw heads pulling through and delamination of paper facer

CONCLUSIONS

Based upon the results of these tests, it was found that the condition of gypsum sheathing board within the EIFS assembly (the gypsum core, as well as the paper facers) was of critical significance in evaluating the EIFS system pertaining to structural integrity, as well as in predicting the anticipated performance of wall panels with respect to resisting stresses arising from design wind pressures on building surfaces. The gypsum sheathing typically represented the "weak link" within each of the samples tested, and system failure occurring (either wholly or in part) due to fasteners pulling through the sheathing accounted for almost 83% of the samples tested. In addition, since each of the test specimens had been constructed utilizing adhered foam insulation board, it would appear that structural integrity of the proprietary, cementitous, adhesive/base coat is not a real concern when installed in accordance with the manufacturer's recommendations.

It is pertinent to point out that the sheathing constitutes the "foundation" of the EIFS wall assembly and if it fails, obviously, the foam and finish system will go with it, regardless of their own adequacy. Therefore, continual maintenance of the sheathing becomes of paramount importance. The relationship between past water damage to the sheathing and the level of wind resistance exhibited by the EIFS was well established by the results of these tests. Unfortunately, an exterior visual survey of the EIFS is incapable of providing reliable information regarding the gypsum sheathing. Likewise, the amount of staining, or the degree of mold and mildew, which might be visible (even from unhindered interior inspections) was not found to be an absolute indicator of the lack of performance actually exhibited by physical testing. Accordingly, it is suspected that some type of field test, similar to the one developed here, will be required in the future in order to evaluate overall wall assemblies.

It is interesting to note that when EIFS was first introduced, most of the manufacturers prohibited the use of mechanical fasteners installed through the insulation layer, favoring instead the adhered construction. Currently, at least one manufacturer markets a fastened only system, while some of the older companies have moved to allow fastening as an "option." In my opinion, utilization of mechanical fasteners through the insulation layer, in which the sheathing is "captured" within the system, provides a more desirable attachment scheme and still retains the economical sheathing as a base substrate. It is anticipated that this type of construction would produce a system in which pull-through of the fasteners would be less likely to take place, since these attachment methods typically use oversized stress plates at the head of the fastener, similar to those utilized in single ply roofing systems.

It is significant to note that, taken overall, about 49% of the samples tested failed to meet the design wind requirements of its particular geographic locale. In the opinion of this author, the results of this test program are not a glowing testimonial for the use of traditional EIFS wall construction using adhered assemblies. It should be remembered, however, that each of these projects had experienced some degree of moisture distress, which prompted the investigations leading to the test program being implemented. In addition, although some of the moisture damage could be attributed to cracking within the EIFS assemblies, other sources of water leakage (such as adjacent windows and perimeter sealant applications) which did not have anything directly to do with the EIFS installation were also present. Accordingly, it could be argued that these projects do not portray a representative sample of current EIFS usage. While this point may have some merit, it should be noted that the projects included within this test program did not, in all cases, consist of workmanship that would be considered sub-standard. In other words, with the possible exception of the South Central state project, the level of workmanship observed appeared to be about normal for the numerous EIFS projects inspected by this author. It could also be granted that the majority of these failures were contributed from one project, which experienced a failure rate of almost 60%, while the other two projects exhibited a failure rate of only 12.5%, when isolated from the particularly bad property. Still, in terms of structural integrity, an 88% reliability is hardly acceptable for buildings having human occupancy.

Based upon the field experience of this author, it is my opinion that appropriate field tests should be formally developed for EIFS wall assemblies in order to evaluate long term system performance. These formal test methods should include criteria for determining the deterioration that may have been experienced by any individual EIFS assembly due to poor workmanship during original construction, failure of allied components of the building envelope (i.e. joint sealants) which affect performance of the EIFS, or physical damage incurred from building use and maintenance (i.e. window washing equipment, etc.).

DISCUSSION

There are several topics of interest, which it is felt need to be addressed with regard to this paper and the test program reported. With regard to the test method utilized for these buildings, it could be argued that cutting through the gypsum sheathing around the adhered wooden test module negates the potentially beneficial effect of adjoining portions of the wall assembly. It could be postulated that the contributing forces would behave in much the same manner as a structural "diaphragm" assembly. This possibility was considered when the test method was being developed; however, the instrumentation and test apparatus required to evaluate this effect was deemed to be too involved for the purpose intended for these tests. In addition, according to applicable codes, each square foot of EIFS material should be capable of resisting applied loads anticipated by design wind criteria. Finally, as has been alluded to, this procedure leans heavily upon previously existing test standards which also isolated the test specimen from the surrounding assemblies. For these reasons, it is felt that the test method utilized here is a valid indicator of the EIFS performance.

Recently, manufacturers of EIFS have begun to promote the "option" of using proprietary gypsum products which utilize glass fiber facers in lieu of paper, and have ostensibly been formulated for weather resistant performance. At this point, it is not clear that use of such a product would exhibit improved results compared to those obtained in this test program. This is so primarily because the predominate location of damage observed at these projects was related to openings and penetrations through the EIFS where the gypsum sheathing had been cut and did not retain a factory edge. In my opinion, this condition would not be significantly changed with the use of weather resistant gypsum boards. Further, this author has inspected several projects using such gypsum products in which the sheathing was allowed to be exposed to the weather for a prolonged period of time prior to installing the wall finish. While these products undeniably exhibited improved short-term weathering characteristics, in each case locations were found at which the glass fiber facer was easily peeled at cut edges, etc. In addition, the long-term effect of such exposure cannot be accurately predicted without more research and data pertaining to job histories.

ACKNOWLEDGEMENTS

The author wishes to express appreciation for the work of several associates who contributed to the development of this paper, as well as those who consented to the use of materials and information accumulated through these efforts. Those included in this group are Messrs. D.B. Hales, Mark C. Fox, William Cowart, and Chris Clarke, all of the firm of Moisture Systems, Inc. Also, thanks goes to Sharon Laird for review and editing of the manuscript.

REFERENCES

[1] Foster, Allan D., "Performance Characteristics of Exterior Plastering Systems," The Construction Specifier, The Construction Specifications Institute, Volume 42, Number 8, August 1989; p. 90.

[2] "Loss Prevention Data 1-52; Field Uplift Tests," Loss Prevention Data for Roofing Contractors, Factory Mutual Engineering Corporation, February 1986.

[3] Doyle, Margaret, "Trends in specifying EIFS," Building Design & Construction, Volume 29, Number 8, August 1988; p. 60.

[4] Russell, James S., "EIF Systems: quality lost in the translation?," Architectural Record, July 1989; p. 124.

[5] "Moisture in Building Construction," in ASHRAE Handbook of Fundamentals, 1985, pp. 21.17-21.19.

Yong-Xin Tao, Robert W. Besant, and Kamiel S. Rezkallah[1]

HEAT AND MOISTURE TRANSPORT THROUGH A GLASS-FIBER SLAB WITH ONE SIDE SUBJECT TO A FREEZING TEMPERATURE

REFERENCE: Yong-Xin, T., Besant, R. W., and Rezkallah K. S., "Heat and Moisture Transport through a Glass-Fiber Slab with One Side Subject to a Freezing Temperature," Water in Exterior Building Walls: Problems and Solutions. ASTM STP 1107, Thomas A. Schwartz, Ed., American Society for Testing and Materials, Philadelphia, 1991.

ABSTRACT: Experiments were performed of heat and moisture transport through a glass-fiber slab to investigate the effect of condensation and frosting on the thermal performance of insulation. The slab has one side impermeable and subject to a cold temperature (below the triple point of water) and the other side open to a moist air at a room temperature and for various humidity levels. The accumulation of moisture and frost results in an increase in heat loss up to a maximum of twice as much as that for a dry insulation. A comparison between a one-dimensional, numerical model based on a formulation of vapor diffusion in porous media and the experimental results, reveals that significant hygroscopic effects occur in the experiment for initially oven-dried specimens, and contribute to the discrepancy between the transient results of the prediction and of the experiments. For the transient processes, the differences in predictions, excluding the adsorption phenomenon, and the experimental data can lead to a maximum discrepancy of 26% in temperature and of 35% in heat flux.

KEYWORDS: heat transfer, moisture transport, fibrous insulation, frost, adsorption.

NOMENCLATURE

C	moisture content, kg/m^3
c_p	heat capacity at constant pressure, J/kg-K
$D^*_{v,eff}$	effective vapor diffusivity, $\varepsilon_\gamma D/\tau$, m^2/s
h_{fg}	enthalpy of vaporization, J/kg
h_{sg}	enthalpy of sublimation, J/kg
k	thermal conductivity, W/m-K
L	characteristic length of the slab, m
$\dot{m}$	dimensionless rate of phase change

---

[1] Dr. Tao is a Post Doctoral Fellow, Professor Besant is the Head of the Dept., and Dr. Rezkallah is an Associate Professor of Department of Mechanical Engineering, University of Saskatchewan, Saskatoon, Sask. S7N 0W0, Canada.

p	pressure, Pa
$\dot{q}''$	heat flux, W/m^2
Q'	heat flux ratio defined in equation (12)
R_v	vapor gas constant, J/kg-K
t	time, s
T	temperature, K
z	coordinate axis, m
$\alpha_{0,eff}^*$	effective thermal diffusivity, $k_{0,eff}^*/\rho_o^* c_o^*$, m^2/s
ΔT	reference temperature difference, $T_a^* - T_c^*$, K
ε	volume fraction
ϕ	relative humidity
ρ	density, kg/m^3

Subscripts

a	air
c	cold
f	frozen region
ref	reference point
w	wet region
β	liquid phase in wet region or ice phase in frozen region
γ	gas phase which consists of air and water vapor
σ	solid phase

INTRODUCTION

Heat and moisture transport through insulation materials has been of considerable interest among researchers, not only because of its practical significance in energy management for building and refrigerated envelopes, but also because of the physical complexity of the problems in various circumstances. Extensive studies have been conducted analytically and numerically [1-3] along with a few experimental efforts [4-6]. Condensation effects in insulations have been investigated rigorously for temperature ranges above the freezing point [7,8]. It is shown that the additional, steady-state heat loss due to condensation of water in an insulation slab under a thermal gradient is negligible if the mass transport process in the slab is dominated by vapor molecular diffusion (i.e., Pe=0) and the Lewis number (α_{eff}/D_{eff}) is small [8]. This conclusion was supported by an early experimental study by Kumaran [6], in which a glass-fiber insulation slab, after being open to a moist air at 97% relative humidity for a long time, showed no significant increase in heat flux as compared to a dry slab. It has also been reported [5] that for an insulation slab with its impermeable cold side (a vapor retarder applied) and the warm side open to a forced convective moist air at a relative humidity of less than 80%, the heat loss is almost the same as if the slab is dry. However, when the cold side of the insulation slab is subject to a temperature below the triple point of water, condensed water may exist as frost which both alters the temperature distribution and increases the effective thermal conductivity of the slab. Therefore, it is expected that the heat loss, compared with that for a dry insulation, would be larger than the case with only condensation effects. A numerical study [9] shows that condensation and frosting in a typical glass-fiber slab with one boundary impermeable, will result in a 30% increase in heat flux at the quasi-steady-state when the ambient air humidity is above 60% and the moisture transport process is vapor-diffusion dominated. This prediction of the increase in heat loss due to frosting effects needs to be confirmed by experiments. Despite the above-mentioned efforts, this literature survey reveals that frosting effects on the thermal performance of insulation have not been studied rigorously. Furthermore, no significant experimental data are available under controllable experimental conditions for the cold temperature below the freezing point.

Motivated by the above observations, we performed experiments in this study to investigate the thermal performance of an insulation slab with a possible accumulation of moisture/frost. The warm side of the slab is open to a moist air flow under various ambient humidity levels. The cold boundary of the slab is impermeable, as if in the case where an ideal vapor barrier is used. This condition is close to the operational conditions of some insulation materials. The imposed cold boundary temperature is below the triple point of water. Besides reporting the experimental data, one of the objectives of this study is to compare the experimental results with the prediction done previously [9], in which it is assumed that the mode for moisture transport in the slab is primarily molecular diffusion; i.e., no convective fluid flow through the insulation is considered, and no hygroscopic effects exist. Although it is found that hygroscopicity may not significantly influence the heat flux in some circumstances [10], our experiments show it is important in transient processes in which the initially dry glass-fiber slab is under a temperature gradient and is continuously open to a moist air.

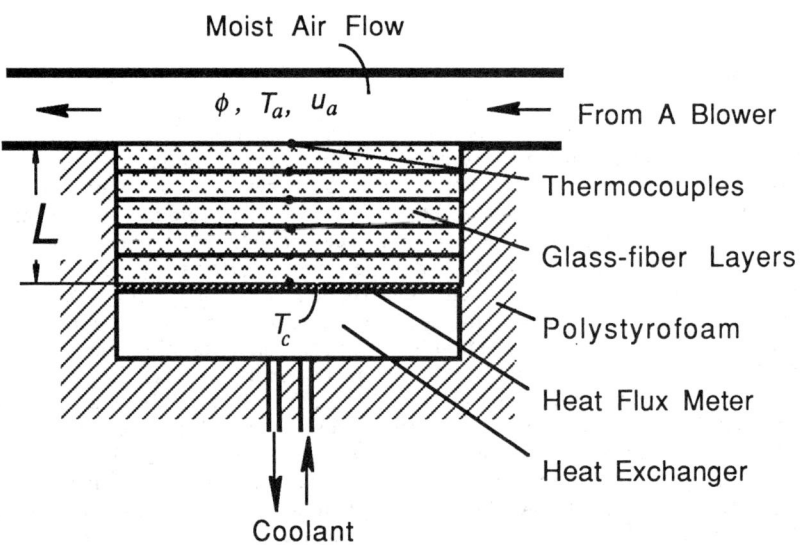

Fig. 1 -- A schematic of the experimental apparatus.

EXPERIMENT

A schematic of the apparatus is shown in Fig. 1. A five-layer glass-fiber slab (AF545, Fibglas Canada, ON) with a total dimension of 280 x 600 x 95 mm, is placed on a cold plate which can be cooled to below the triple point temperature of water through a heat exchanger. Ethylene glycol-water solution is used as a coolant and is supplied by a pump from a storage tank which is placed in an existing, environmental chamber. A heat flux meter is sandwiched between the slab and the top surface of the heat exchanger to allow for the measurement of the heat flux leaving the slab. The heat flux meter consists of a polyethylene sheet with a 1/8 inch (3.175 mm) thickness and 21 thermocouples at each side of the sheet. An aluminum sheet with a 1/8 inch (3.175 mm) thickness is mounted on the top of the sheet to keep the temperature constant. The thermal conductance (W/m^2 °C) of the heat flux meter is calibrated *in situ* using a heat flow transducer (provided by National Research Council-IRC, Saskatoon). Another 12 thermocouples are placed between the glass-fiber layers to allow for recording the temperature field of the slab. The upper surface of the slab is open to a fully developed turbulent air flow, and the other sides of the slab are covered by plastic sheets. The air temperature and relative humidity can be controlled and are measured by thermocouples and humidity sensors, respectively. The air velocity is measured through an orifice meter in the downstream of the channel. The edges of the glass-fiber slab are insulated by polystyrofoam boards to minimize the edge heat loss.

The glass-fiber slab is initially dried in an oven at 105 °C for about 15 hours, and then wrapped by a plastic sheet and cooled to the room temperature. After the slab is properly mounted in the apparatus, moist air in the duct and the coolant in the heat exchanger are supplied at the same time. The air temperature and humidity and the cold temperature are stabilized within 10 to 15 minutes. The temperatures of the glass-fiber slab, air and cold plate are recorded and monitored by a personal computer through a data acquisition unit. A typical experiment runs about 3 to 4 hours which covers both a transient and a quasi-steady-state period. At a desired time, the slab is taken out of the apparatus and each layer is weighed using an electronic scale, to find a total amount of moisture/frost accumulation during that period of time. For the purpose of comparison, we also perform the experiments for dry specimens in which the slab is completely wrapped with plastic sheets and the test is run at a low ambient humidity for the same air flow rate and air temperature.

From the experiments, the time variation of the temperature field in the slab, the time variation of the heat flux at the cold side of the slab, and the moisture/frost accumulation can be found for various air velocities, temperatures, relative humidities, and cold-plate temperatures. The uncertainty in the heat flux measurement is estimated within 6%. All thermocouples are calibrated using a standard calibrator and the uncertainty in temperature measurement is within 0.1 °C. The relative humidity is within 2% and the electronic scale is accurate to 0.1 gram. Table1 gives the range of experimental conditions and the basic properties of the glass-fiber slab used.

TABLE 1 -- The Experimental conditions and the Properties of the Glass-fiber Slab

T_a	°C	20 ~ 25
T_c	°C	−20 ~ −6
ϕ		0.4 ~ 0.95
u_a	m/s	4.6
h_a	W/m^2-K	25
ϵ		0.97
ρ	kg/m^3	67.0
$k_{eff,0}$	W/m-K	0.055
L	m	0.09509 (0.75 inch × 5)

TABLE 2 -- Dimensionless Variables

ρ	c_p	T	ρ_i	k_i	p_i	z	c_i	k_{eff}	t	$\dot{m}$
$\dfrac{\rho^*}{\rho_0^*}$	$\dfrac{c_p^*}{c_0^*}$	$\dfrac{T^*}{\Delta T^*}$	$\dfrac{\rho_i^*}{\rho_0^*}$	$\dfrac{k_i^*}{k_{0,eff}^*}$	$\dfrac{p_i^*}{p_{v,0}^*}$	$\dfrac{z^*}{L}$	$\dfrac{c_i^*}{c_0^*}$	$\dfrac{k_{eff}^*}{k_{0,eff}^*}$	$\dfrac{t^*}{L^2/\alpha_{0,eff}^*}$	$\dfrac{\dot{m}^*}{\rho_0^*\alpha_{0,eff}^*/L^2}$

TABLE 3 -- Dimensionless Parameters

Ψ_D	P_1	P_4	P_8	P_{10}	P_4'	P_{10}'
$\dfrac{D_{v,eff}^*}{\alpha_{0,eff}^*}$	$\dfrac{\rho_\beta^*}{\rho_0^*}$	$\dfrac{h_{fg}^*}{c_0^*\Delta T^*}$	$\dfrac{\Delta T^* R_v^* \rho_0}{p_{v,0}^*}$	$\dfrac{h_{fg}^*}{R_v^*\Delta T^*}$	$\dfrac{h_{sg}^*}{c_0^*\Delta T}$	$\dfrac{h_{sg}^*}{R_v^*\Delta T^*}$

ANALYSIS

The following one-dimensional, transient analysis of simultaneous heat and mass transfer in insulation with phase changes is developed [9] in order to predict the distributions of temperature, vapor density and moisture/frost content (in terms of the volume fractions of liquid and frost) in an insulation material and the variation of heat flux with time, and to compare with the experimental results. The major assumptions are (a) there is no hygroscopic effects (excluding adsorption processes), (b) all phases in the porous medium are in local thermodynamic equilibrium, (c) accumulated frost does not exist as a self-porous medium (i.e., only isolated ice crystals are considered), and (d) a dry, wet, and frozen region may coexist in the slab. The schematic of this model is depicted in Fig. 2. For this model, it is obvious that any moisture that might be accumulated in the insulation is transported from the ambient air through vapor diffusion under the influence of a thermal gradient. The other assumptions and derivations of this model are referred to in [9]. Here we only present the governing and constitutive equations.

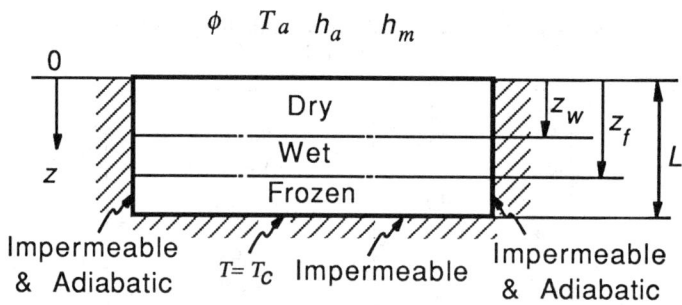

Fig. 2 -- Transient condensation and frosting in an insulation slab with prescribed boundary conditions.

Governing equations

The coupled, non-linear partial differential equations below (non-dimensionalized according to the definitions listed in Tables 2 and 3) are used to describe the transport phenomena.

Liquid/ice phase continuity equation:

$$\frac{\partial \varepsilon_\beta}{\partial t} + \frac{\dot{m}}{P_1} = 0 \tag{1}$$

Gas diffusion equation:

$$\frac{\partial(\varepsilon_\gamma \rho_v)}{\partial t} - \dot{m} = \frac{\partial}{\partial z}\left(\Psi_D \frac{\partial \rho_v}{\partial z}\right) \tag{2}$$

Energy equation:

$$\rho c_p \frac{\partial T}{\partial t} + \dot{m} P_4 = \frac{\partial}{\partial z}\left(k_{eff} \frac{\partial T}{\partial z}\right) \tag{3}$$

Where P_1 in equation (1) and P_4 in equation (3) may be replaced by P_1' and P_4', respectively, for the frozen region (see Table 3), and $\dot{m}$ is the dimensionless mass rate of phase change per unit volume.

Additional equations used are:

Volumetric constraint:

$$\varepsilon_\sigma + \varepsilon_\beta + \varepsilon_\gamma = 1 \tag{4}$$

Thermodynamic relations:

$$p_a = p_t - p_v \tag{5}$$
$$p_a = P_{11}\rho_a T \tag{6}$$
$$p_v = P_s \rho_v T \tag{7}$$

and for saturation conditions

$$p_v = \exp\left[-P_{10}\left(\frac{1}{T} - \frac{1}{T_{ref}}\right)\right] \tag{8}$$

where

$$\rho = \varepsilon_\sigma \rho_\sigma + \varepsilon_\beta \rho_\beta + \varepsilon_\gamma(\rho_v + \rho_a) \tag{9}$$

$$c_p = \frac{\varepsilon_\sigma \rho_\sigma c_\sigma + \varepsilon_\beta \rho_\beta c_\beta + \varepsilon_\gamma(c_v \rho_v + c_\alpha \rho_\alpha)}{\rho} \tag{10}$$

$$k_{eff} = \varepsilon_\sigma k_\sigma + \varepsilon_\beta k_\beta + \varepsilon_\gamma \frac{k_v \rho_v + k_\alpha \rho_\alpha}{\rho_v + \rho_\alpha} \tag{11}$$

Also, in equation (10), P_{10} may be replaced by P'_{10} for the frozen region.

In order to examine the thermal performance of the insulation slab, the following heat flux ratio is defined

$$Q' = \frac{Q_c}{Q_{c0}} = \frac{\left(-k_{eff}\frac{\partial T}{\partial z}\right)_c}{\left(-k_{eff}\frac{\partial T}{\partial z}\right)_{c0}} \quad (12)$$

where Q_c is the heat flux leaving the cold boundary and Q_{c0} is the heat flux leaving the cold boundary of the same insulation slab subject to the same boundary conditions except ε_β is equal to zero at any time and location, i.e., the entire slab remains dry.

The boundary and initial conditions are depicted in Fig. 2. The finite difference forms of equations (1) to (3) are derived using the implicit scheme with the backward difference for the time derivative. Again, readers may refer to [9] for the detailed discussion on computational schemes.

RESULTS AND DISCUSSION

Experimental results

In Fig. 3, a typical example of the measured temperature and heat flux history is shown. Due to the thermal capacitance of the heat exchanger, the cold plate temperature becomes stable about 5 minutes later after the initial time. This time lag is much shorter than the time required for the insulation to reach a quasi-steady state (about 2 hours). As expected, the thermal response of the slab is not significantly influenced by the response of the cold-plate heat exchanger.

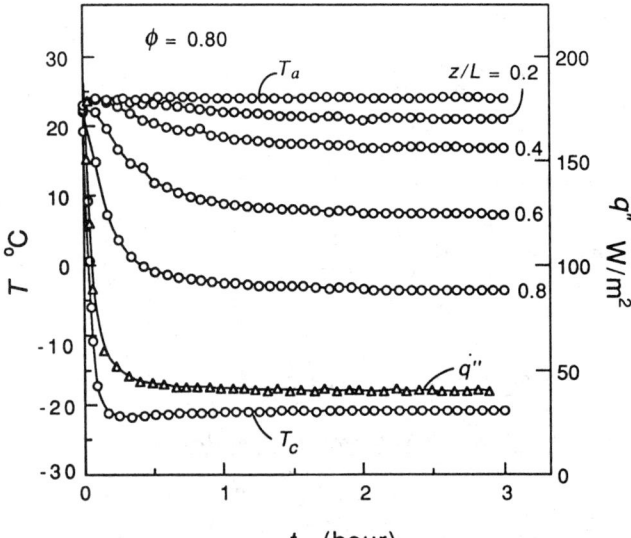

Fig. 3 -- A typical example of the time variation of the measured temperatures at various locations in the slab (initially oven-dried) and the heat flux.

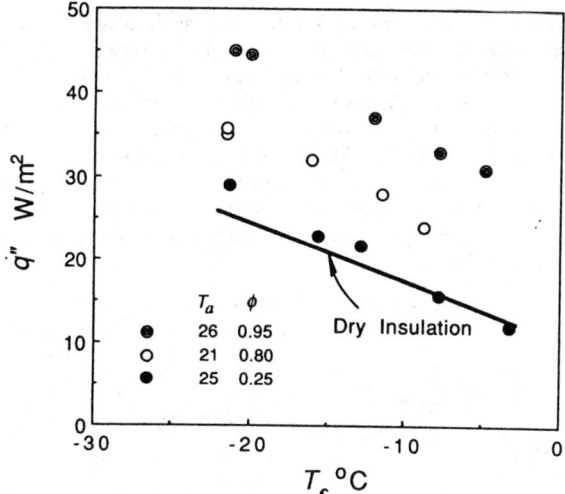

Fig. 4 -- The measured quasi-steady-state heat flux for various boundary conditions: $t = 2.5$ hours.

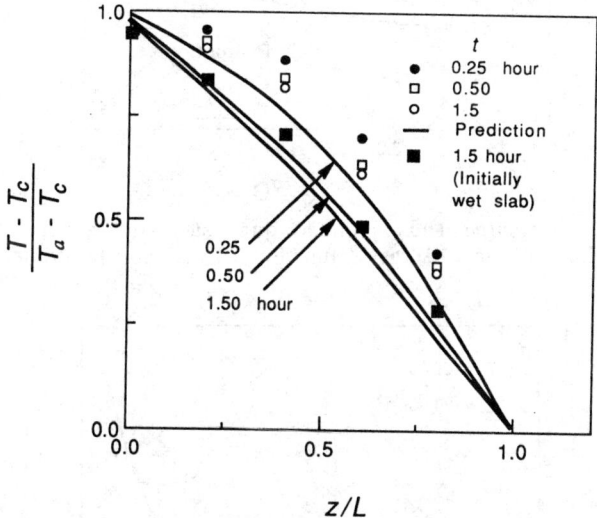

Fig. 5 -- A typical temperature distribution in the slab: $\phi = 0.804$, $T_a = 25\ °C$ and $T_c = -20\ °C$.

In Fig. 4, the quasi-steady-state heat flux for different conditions are shown. As the ambient air relative humidity increases, the heat loss increases. This is because the moisture and frost accumulation increases as the ambient relative humidity increases. Since vapor diffusion is the major mode in moisture transport, the high relative humidity of air means a large partial pressure of water vapor which yields a high vapor diffusion flux towards the cold side of the insulation, and causes an increase in the moisture content in the wet region or frost content in the frozen region. Therefore, the effective thermal conductivity of the insulation is also increased. In addition, condensation and frosting continue to occur even in the quasi-steady state, and the heat released during condensation and ablimation alters the temperature field in such a way that it results in a larger heat flux leaving the cold side of the insulation than that for a ideal, dry specimen. This can be seen in Fig. 5 where the transient temperature distribution at various locations is shown to be convex, or in Fig. 6, in terms of Q', defined in equation (12), is greater than 1. It is also shown in Fig. 7 that Q' increases as the air relative humidity increases and decreases as the cold temperature decreases at high air humidity levels. This trend is consistent with the prediction; although, in terms of the absolute magnitude, the prediction underestimates the results which will be discussed later.

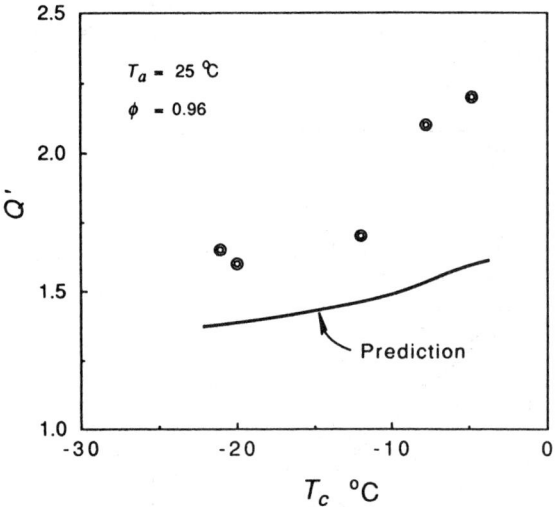

Fig. 6 -- The measured and predicted quasi-steady-state heat flux ratio (defined in Eq. 12) as a function of the cold plate temperature.

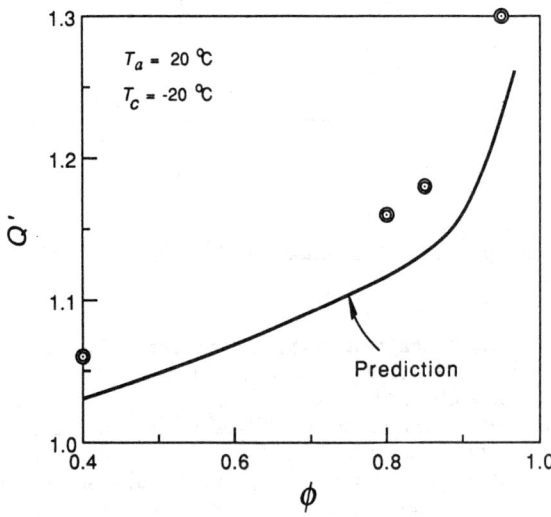

Fig. 7 -- The measured and predicted quasi-steady-state heat flux ratio (defined in Eq. 12) as a function of the ambient air relative humidity.

The measured average moisture/frost accumulation for each layer of the insulation is shown in Fig. 8, where C is defined as the mass of water or frost per unit volume of the insulation slab. The largest accumulation occurs in the layer attached to the cold plate. As observed in the experiments, this maximum frost accumulation is confined to the impermeable surface of the slab (i.e., between the cold vapor retarder and the slab) which is in contact with the cold plate. This trend is generally expected.

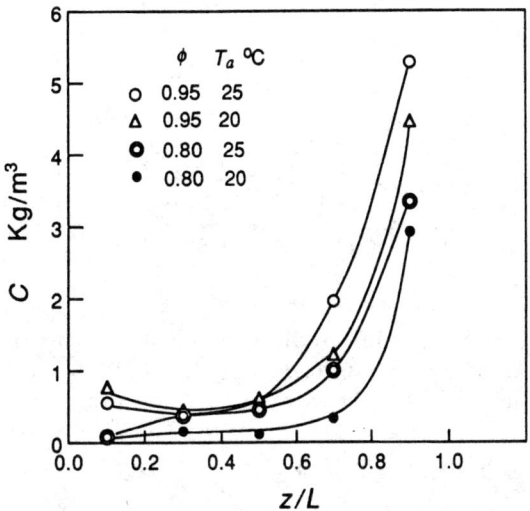

Fig. 8 -- The measured, average moisture/frost accumulation after 3 hours: T_c = -20 °C (the experimental data are the averaged values over each layer of the slab).

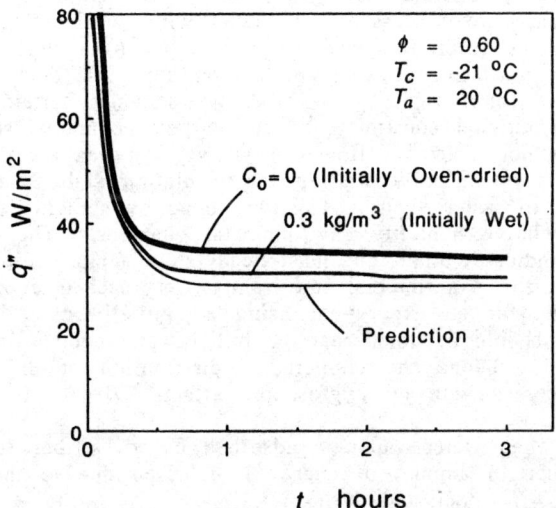

Fig. 9 -- The effects of hygroscopicity on the heat flux measurement.

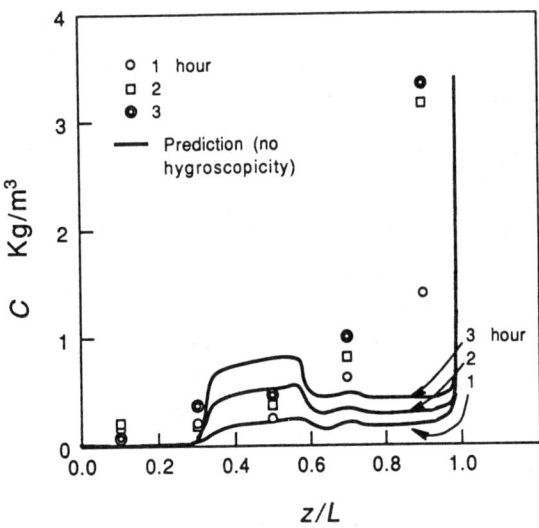

Fig.10 -- The difference in moisture/frost accumulation distribution due to hygroscopic effects: $\phi = 0.80$, $T_a = 25$ °C, and $T_c = -21$ °C (the experimental data are the averaged values over each layer of the slab).

Comparison of data with the predictions

The prediction, based on the model briefly presented in section 3, is also shown in Figs. 5-7. The predicted temperature inside the slab, for a given progressive time, is generally lower than the corresponding measured one. This temperature distribution also contributes to the underestimate of the quasi-steady-state heat flux. This discrepancy between the measured data and the numerical model is believed to be due to the hygroscopic effects which, for an initially oven-dried glass-fiber slab, leads to the adsorption of water vapor. The adsorption of vapor over glass fibers coated with the bonding material can cause a long time delay before a quasi-steady-state, which is close to the predicted one, can be reached. It was reported that this time delay could be as long as 20--40 hours for a typical glass-fiber board under certain conditions [6]. Due to the operational constraints of our current apparatus, such a long period of the test is not realized. However, we verified this argument about the hygroscopicity effects by initially wetting the insulation slab to minimize the additional adsorption of water vapor. This was done by spraying the slab with steam in a 100% relative humidity environmental chamber. The results, shown in Fig. 5, indicates that the quasi-steady-state temperature distribution, as predicted (within the experimental uncertainty), is reached at a time which is much shorter than for an experiment using an initially oven-dried slab (also see Fig.9). It should be noted that the initial water content in the slab does not significantly change the temperature distribution at the quasi-steady-state for processes with no hygroscopic effects [9].

Physically, when water vapor is adsorbed on glass-fiber surfaces of a dry insulation, a certain amount of energy is released due to the interaction between vapor molecules and solid (fiber) surfaces. According to thermodynamics, the energy exchanged in an adsorption process (per unit mass of adsorbate) is generally defined as heat of adsorption which, depending on the process involved, can be defined as either isothermal or adiabatic. The heat of adsorption is generally a function of temperature and vapor concentration, and can be up to four times as high as the heat of condensation for many hygroscopic materials [11]. This means that when the adsorption phenomenon contributes to the overall heat and mass transfer in an initially dry insulation, the heat of adsorption must be included in the computational formulation, replacing the heat of condensation (see Eq. 3, the parameter P_4). As a consequence of this heat of adsorption (analogous to the latent heat), the temperature distribution in the slab shows a convex shape, as depicted in Fig. 5. For a non-hygroscopic material, no adsorption process would occur before bulk condensation occurs (dry period) and, therefore, the energy equation (3) excludes the latent heat term, which leads to a lower temperature prediction than for a hygroscopic material. Since moisture transport is strongly coupled with heat transfer, the difference in the temperature fields between the process with the hygroscopic effect and that without it, will result in a difference in moist/frost accumulation, as shown in Fig.10. In addition, the amount of vapor adsorption is also a function of temperature and vapor concentration, and increases as temperature decreases. This indicates that a hygroscopic material that is in adsorption equilibrium at a room temperature, can still adsorb moisture when its temperature is decreased. This is especially noticeable for the insulation slabs in low temperature applications, as shown from our experimental results. Also, in the transient problem considered here, the continuous vapor mass transfer from the ambient to the insulation yields that there exists no strict steady-states; therefore, the results can only be compared at a fixed time.

SUMMARY

Experiments were performed to investigate the effects of moisture/frost accumulation on thermal performance of a glass-fiber insulation slab which is subject to a moist, warm air flow. The cold temperature is varied from -5 °C to -21 °C, and the air humidity changes from 40% to 95%. In this range, it is found that the maximum increase in the heat loss from the cold side for a typical slab, can be 100% greater than that for the same slab under the dry condition. The presented, one-dimensional, vapor diffusion model predicts the simultaneous heat and mass transfer in insulation with no hygroscopic effects. The experimental observation indicates that the hygroscopic effects are significant for initially oven-dried specimens. For the transient processes, the differences in predictions, excluding the adsorption phenomenon, and the experimental data can lead to a maximum discrepancy of 26% in temperature and of 35% in heat flux. We conclude that the adsorption effects are important in studying the transient problems of simultaneous heat and moisture transfer in porous insulation. Further study is suggested that will include the hygroscopic effects in the numerical prediction. Also, more experimental data need to be collected to find the hygroscopic properties of porous insulation materials and a practical correlation for heat loss due to moisture/frost accumulation. This information can be used to more rigorously define the testing procedures for those materials so they can be incorporated in the design procedures of thermal insulation for low temperature applications.

ACKNOWLEDGEMENT

The authors would like to thank the Institute for Research in Construction, NRC Canada, for suggesting this problem and for their assistance.

REFERENCES

[1] Eckert, E. R. G. and Faghri, M., "A general analysis of Moisture Migration Caused by Temperature Differences in an Unsaturated Porous Medium," *Int. J. Heat Mass Transfer*, 23, pp.1613-1623, 1980.

[2] de Vries, D. A., "The Theory of Heat and Moisture Transfer in Porous Media Revisited," *Int. J. Heat Mass Transfer*, 30, pp.1343-1350, 1987.

[3] Shapiro, A. P. and Motakef, S., "Unsteady Heat and Mass Transfer with Phase Change in Porous Slabs: Analytical Solutions and Experimental Results," *Int. J. Heat Mass Transfer*, 33, pp.163-173, 1990.

[4] Langlais, C., Hyrien, M. and Karlsfeld, S., "Moisture Migration in Fibrous Insulating Material under the Influence of a Thermal Gradient," *Moisture Migration in Buildings*, ASTM STP 779, pp.191-206, 1982.

[5] Wijeysundera, N. E., Hawlader, M. N. A. and Tan, Y. T., "Water Vapor Diffusion and Condensation in Fibrous Insulations," *Int. J. Heat Mass Transfer*, Vol. 32, 10, pp. 1865-1878, 1990.

[6] Kumaran, M. K., "Moisture Transport through Glass-Fibre Insulation in the Presence of a Thermal Gradient," *Journal of Thermal Insulation*, Vol. 10, pp. 243-255, 1987.

[7] Vafai, K. and Whitaker, S., "Simultaneous Heat and Mass Transfer Accompanied by Phase Change in Porous Insulation," *J. Heat Transfer*, Vol. 108, pp.132-140, 1986.

[8] Vafai, K. and Sarker, S., "Condensation Effects in a Fibrous Insulation Slab," *J. Heat Transfer*, Vol. 108, pp. 667-675, 1986.

[9] Tao, Y.-X., Besant, R. W. and Rezkallah, K. S., "Unsteady Heat and Mass Transfer with Phase Change in an Insulation Slab: Frosting Effects," *Int. J. Heat Mass Transfer*, in press, 1990.

[10] Mitalas, G. P. and Kumaran, M. K., "Simultaneous Heat and Moisture Transport through Glass Fiber Insulation: An Investigation of the Effect of Hygroscopicity", ASME SED. Vol. 4, 1-4, 1987.

[11] Babbitt, J. D., "On the Adsorption of Water Vapor by Cellulose", *Canadian Journal of Research*, Vol. 20, 144-172, 1942.

Harkaran S. Jhinger, Gren K. Yuill, and Tom Hamlin

COMPARISON OF A WALL MOISTURE MODEL WITH FIELD DATA

REFERENCE: Jhinger, H. S., Yuill, G. K., and Hamlin, T., "Comparison of a Wall Moisture Model with Field Data," Water in Exterior Building Walls: Problems and Solutions, ASTM STP 1107, Thomas A. Schwartz, Ed., American Society for Testing and Materials, Philadelphia, 1991.

ABSTRACT: The moisture data measured in several different wall panel configurations at the Atlantic Canada Test Hut Project was analyzed in order to establish a relationship between the panel drying rate and the panel configuration. The permeability of the panel assemblies was found to be a key factor in determining the drying rates of the panels. The panels with higher permeability dried faster than the panels with lower permeability. A dynamic wall moisture simulation computer program, WALLDRY, developed by the Canada Mortgage and Housing Corporation (CMHC) was used to simulate the performance of the panels tested at the Atlantic Canada Hut Project. The results obtained from the WALLDRY simulations were compared with the moisture data measured for the corresponding panel configurations at the test hut project. In about half the panel simulations, WALLDRY adequately predicted the moisture content profiles, however, a number of weaknesses in the model were recognized and modifications to WALLDRY were suggested.

KEYWORDS: moisture, moisture content, wetting, drying, moisture in walls, mass transfer, moisture transfer, heat and moisture transfer, adsorption, moisture simulation, modeling, properties, building, envelope, exterior walls.

Harkaran Jhinger and Dr. Yuill are consulting engineers at G.K. Yuill & Associates Ltd., 200-1200 Pembina Highway, Winnipeg, Manitoba, Canada, R3T2A7; Tom Hamlin is a research scientist with Canada Mortgage and Housing Corporation, 682 Montreal Road, Ottawa, Ontario, Canada, K1A0P7.

INTRODUCTION

The monitoring and collection of the moisture data from the Atlantic Canada Test Huts was carried out by Oboe Engineering Ltd. for the Canada Mortgage and Housing Corporation [1]. The moisture data was collected hourly from the test hut sites in Fredericton, New Brunswick; Halifax, Nova Scotia; and St. John's, Newfoundland. The data monitoring period was from March 1986 to August 1987. In each test hut there were eight test panels in the north wall, all with different material configurations. Similarly, there were eight test panels in the south wall with the same material configurations as in the north wall. The temperature and the relative humidity inside each hut was maintained constant throughout the monitoring period.

The first object of this study was to analyze the measured moisture data and to establish a relationship between the panel configuration and its drying time. A panel or a wall is said to have dried when the moisture content of the framing lumber within the wall has reached below 19%. The method adopted for the analysis was simply plotting the noon-hour stud moisture content values as a function of time and estimating the panel drying times from the graphs.

The second object of the study involved the comparison of the WALLDRY predictions with the measured moisture data. The preliminary WALLDRY simulation of the Atlantic Canada Test Huts indicated that there was not a good match between the WALLDRY predictions and the measured data. A number of variations in the WALLDRY input were tried in order to determine their effect on the WALLDRY results and whether or not they reduce the deviation between the simulation results and the measured data. The permeability of the studs and the Equivalent Moisture Content equation for wood were found to be the only variables that had a significant effect on the moisture content profiles of the studs as predicted by WALLDRY. The Walldry model was calibrated by adjusting these two variables to match the simulations with the measured data from one of the panels. The calibrated model was run to simulate the remaining panels in the three houses.

THE ATLANTIC CANADA MOISTURE TEST HUT DATA

The moisture content of the studs in each panel was monitored and recorded hourly. Other parameters such as temperature, pressure, condensation and relative humidity were also monitored hourly. The location of the sensors installed in these panels is shown in Figure 2, panel 6. The configuration and the construction materials used in each of the eight panel assemblies are illustrated in Figures 1 and 2. Each panel had vinyl siding on the exterior and, 4 mil poly and 12.7 mm GYPSUM on the interior. Other key materials in each panel were:

    Panel 1 - waferboard sheathing, no furring
    Panel 2 - waferboard sheathing, furred siding

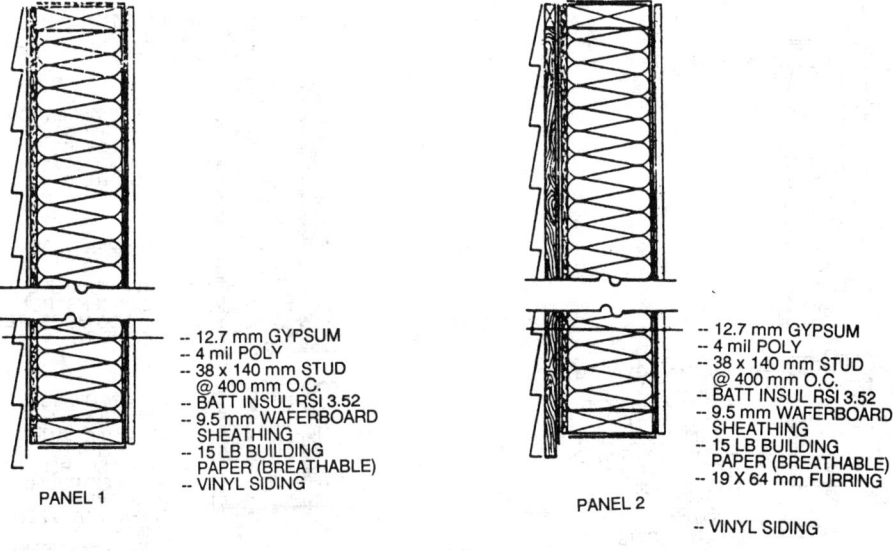
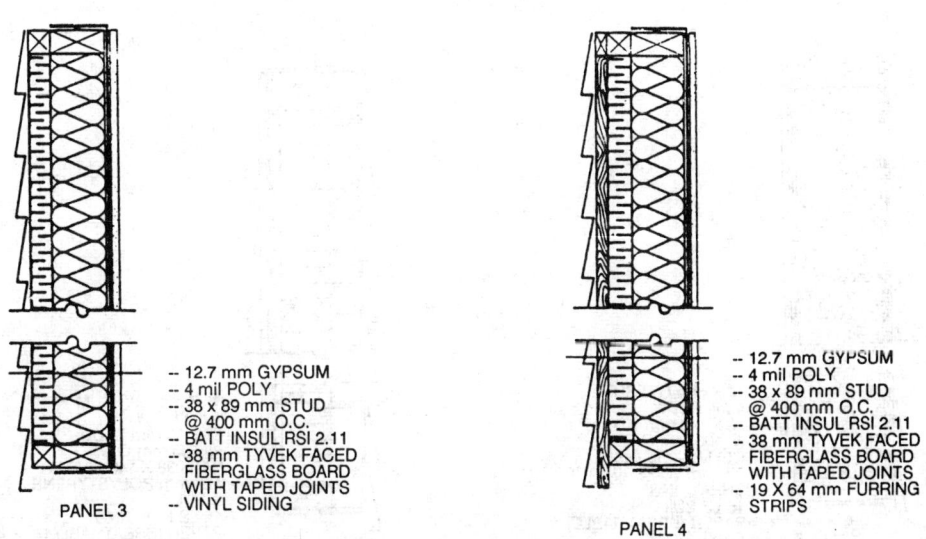

Figure 1 Wall Configurations, Test Panels 1 to 4

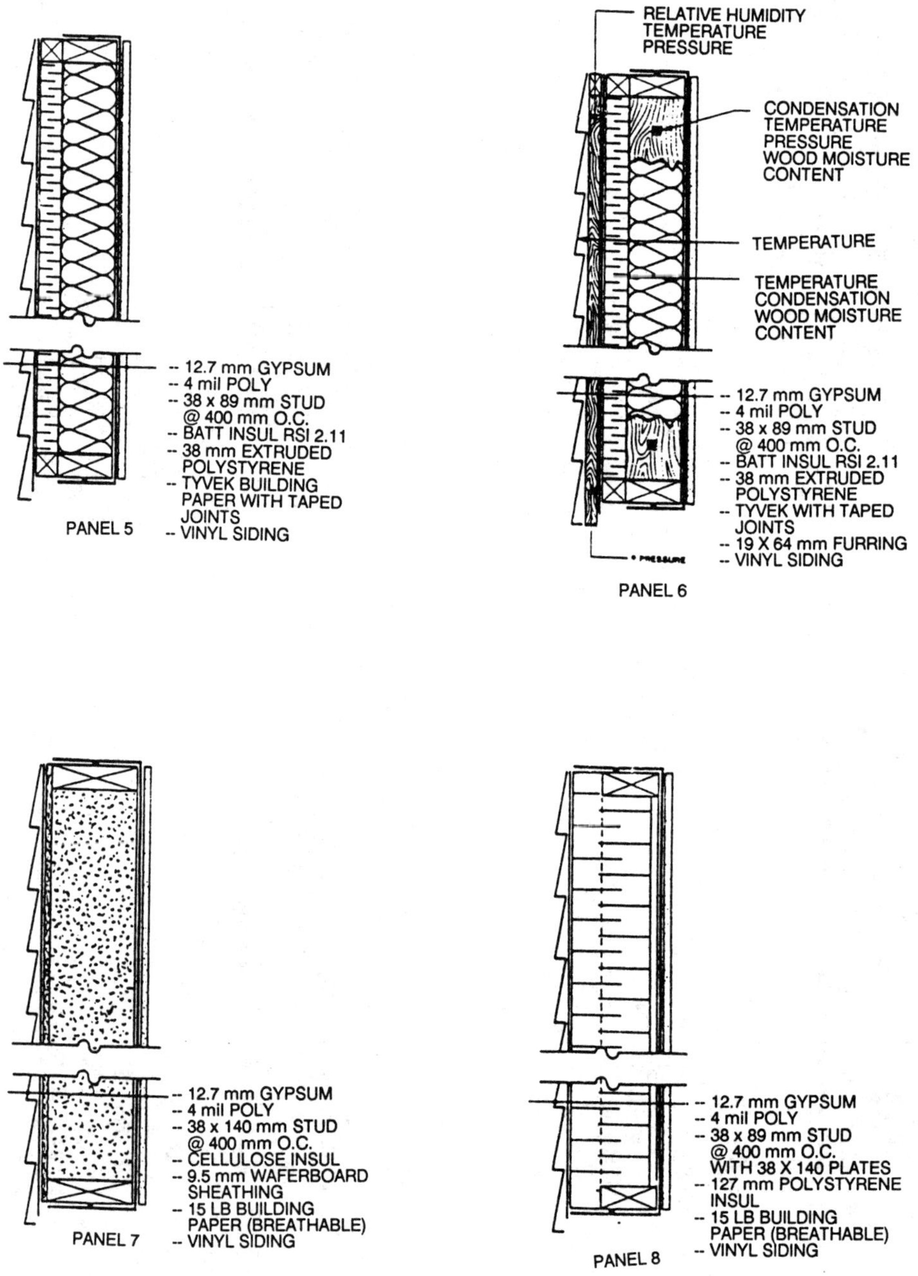

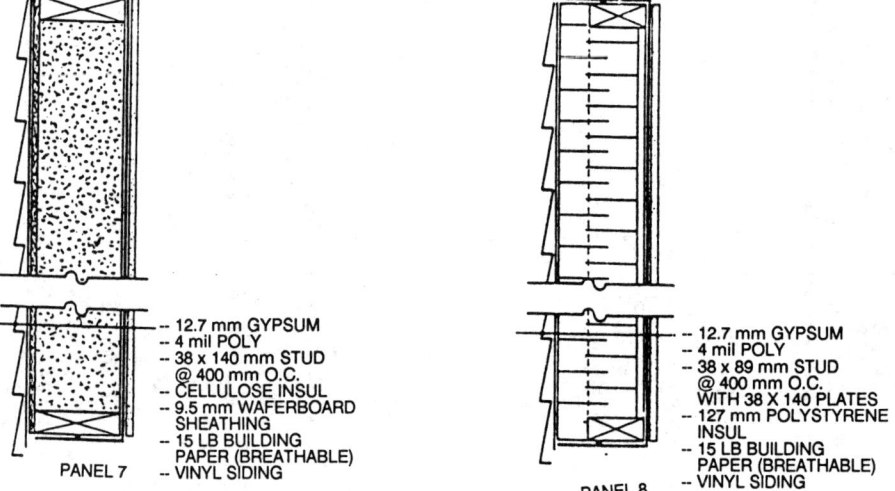

Figure 2 Wall Configurations, Test Panels 5 to 8

Panel 3 - fiberglass sheathing, no furring
Panel 4 - fiberglass sheathing, furred siding
Panel 5 - polystyrene sheathing, no furring
Panel 6 - polystyrene sheathing, furred siding
Panel 7 - wet sprayed cellulose insulation
Panel 8 - polystyrene insulation, sheathing paper, no furring

Use of Noon-Hour Data for the Analysis

A preliminary analysis of the corrected moisture data revealed that the diurnal fluctuations were only significant during the first month of data collection (Figure 3). It was suspected that the large fluctuations were due to the sharp temperature changes during the initial stages of the monitoring period. Therefore, an analysis based only on the noon-hour values was used. Since the magnitude of the Atlantic Canada Moisture data was so great, the data had to be simplified before any statistical or graphical analysis could be carried out. The data was simplified by extracting the noon hour values for the purpose of the analysis. The simplified data files, being much smaller in size, were easily graphable. This approach obscured the short-term (hourly) wetting and drying processes, but analyzing these short-term processes provides little information about the long-term drying rates which are influenced by wall configurations and materials.

Analysis of the Moisture Data

The purpose of the analysis was to establish a relationship between the time a panel takes to dry to 19% and the configuration of the panel. The moisture content of the studs was measured in all panels in the three test huts monitored. The moisture content of the panel stud was plotted against time in order to obtain the panel drying time. An example of such a graph is shown in Figure 4. The drying time for both panel 3 and panel 4 in this figure is about 12 weeks. The average drying times for the three houses in the first phase are summarized in Table 1. On the average, the furred panels 2, 4 and 6 dried slightly faster than the non-furred panels 1, 3 and 5 respectively. The difference in the drying times between the furred and non-furred panels was minor in most cases. However, the panels with high permeability showed a considerable difference in the drying times of the furred and non-furred panels. On the average panel 3 dried in 19 weeks and its furred version, panel 4 dried in 7 weeks. The effect of furring strips on the drying time of lower permeability panels was negligible. This result is not surprising. There are two main resistances to moisture flow from the surface of the studs to the atmosphere. One is the sheathing and the other is the siding. If the sheathing is impermeable, its resistance predominates, and enhancing the removal of moisture from the outside of the sheathing by the use of furring strips has a negligible effect. If, on the other hand, the permeability of the sheathing is high, then the resistance of the siding to moisture transport becomes the predominant effect. In this case, the use of furring strips decreases this critical resistance, and the wall dries significantly faster. This suggested that a

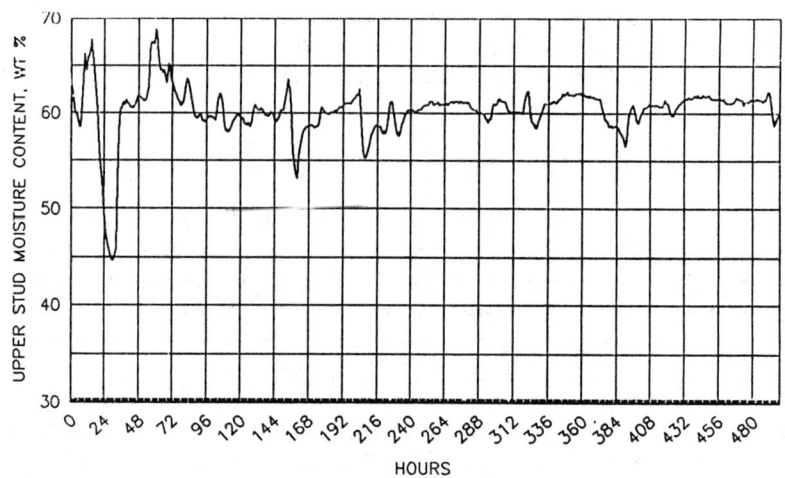

Figure 3

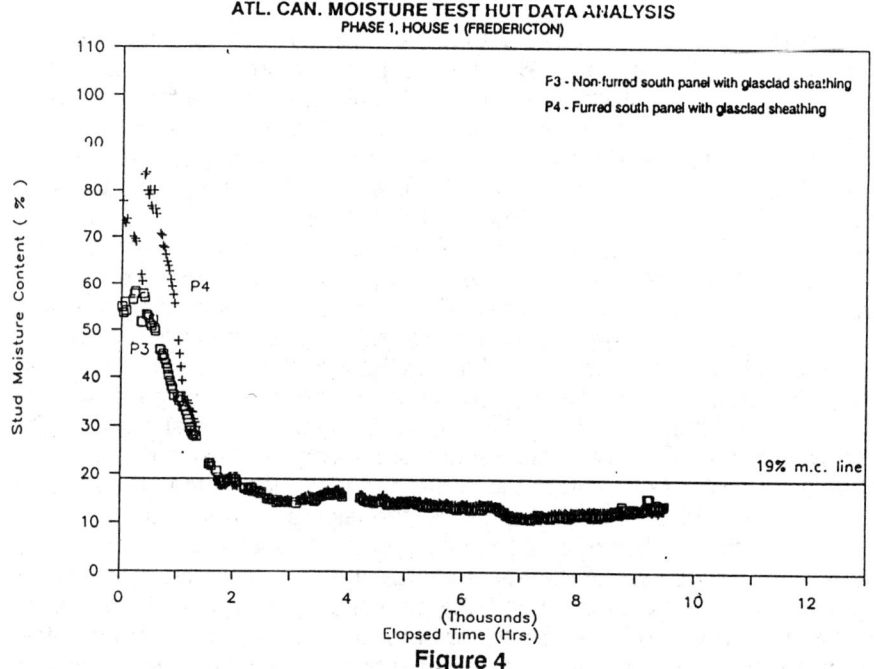

Figure 4

TABLE 1-- Drying Times, ELA and Permeability For South Panels

Panel No.	Drying Time (Weeks)			Average Drying Time (H1+H2+H3)/3 (Weeks)	ELA cm^2	Permeability of the Sheathing Assemblies $ng/Pa.s.m^2$
	House 1 (H1)	House 2 (H2)	House 3 (H3)			
1	36	49	42	42	1.50	43
2	17	64	42	41	0.90	43
3	11	36	11	19	0.96	5366
4	11	08	02	07	1.10	5366
5	42	68	45	52	0.65	35
6	**	60	34	47*	0.85	35
7	**	**	**	**	0.88	43

*AVERAGE BASED ON LESS THAN THREE HOUSES. **THE PANEL NEVER DRIED

combination of furring strips and permeable sheathing should be used wherever wet framing lumber is likely to be used and it is necessary to ensure that it can dry quickly.

The air-tightness and permeability data [2] in Table 1 was used to produce Figures 5 and 6, which demonstrate the effect of these two parameters on the average drying times (of the three huts) of the various panel configurations. Since both these relationships would be expected to be inverse ones, hyperbolas were fitted to the data. The average drying time for each panel was calculated based only on the panels in the south walls. For example, the average drying time for panel 1 in the south wall of all the three huts is 42 weeks (Table 1, column 5). As would be expected, Figure 5 showed that the higher the air-tightness of the panel (that is, the lower the Equivalent Leakage Area, ELA) the longer it takes it to dry. Figure 6 showed that the higher the permeability (or lower the diffusion resistance) the shorter the drying time. When the average drying times were plotted against permeability, the panels with approximately the same ELA (panels 2, 3, 4, 5 and 6) were used; and when the average drying times were plotted against ELA, the panels with approximately the same permeability (panels 1, 2, 5 and 6) were used. In physical terms, this means that, provided the moisture migration potential exists in that direction, the migration of moisture from the wall cavity outwards is highly dependent on the permeability and the air-tightness of the wall sections. Panels 3 and 4 have the highest permeability and thus have the lowest average drying times (19 and 7 weeks respectively), whereas panels 5 and 6 with the lowest permeability result in the highest average drying times of 52 and 47 weeks respectively.

It was apparent from the analysis that the orientation of the panels affects their drying rates. This effect was not obvious from an inspection of the

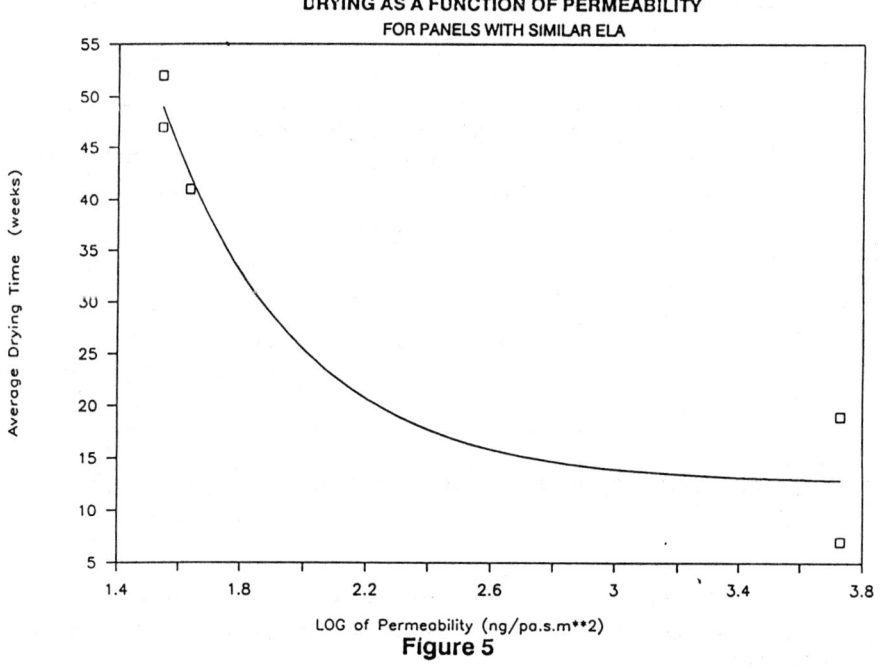

Figure 5

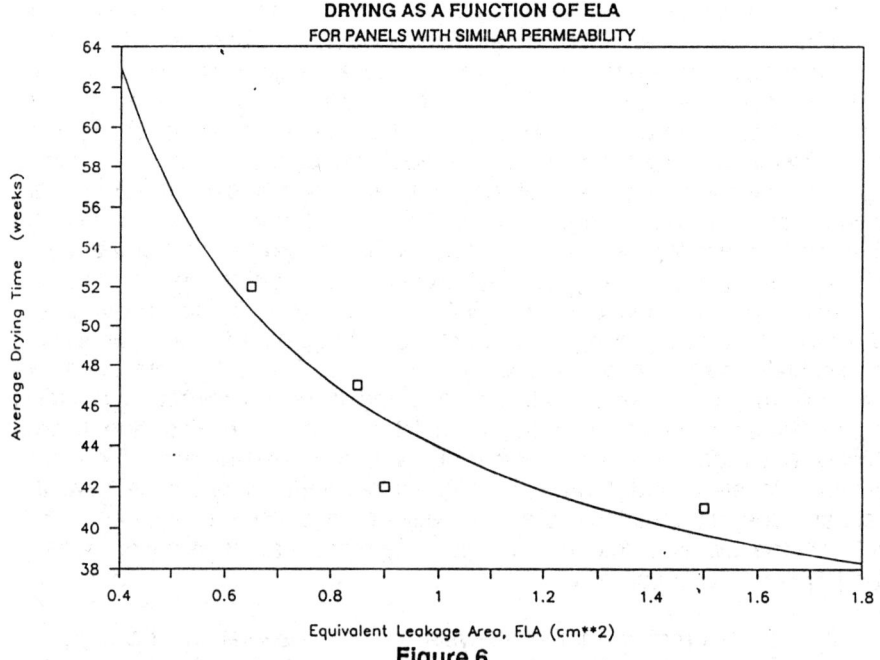

Figure 6

shapes of the drying curves, but it was evident from the times to dry to 19%. The panels facing the south were found to dry faster than the panels facing the north.

The measured noon-hour ambient data from the test hut in Fredericton was plotted (Figures 7 to 10) to evaluate the temperature and pressure differences across the north wall and across the south wall. The temperature and pressure differentials across the wall were not measured separately for each panel. There was only one common measurement for all the panels in one wall. Figures 7 and 8 showed that the noon hour pressure differential profiles for the north wall and the south wall were nearly the same. However, Figures 9 and 10 showed that the noon hour temperature differential profiles for the north wall and the south wall were substantially different. The north wall showed larger temperature differentials across it than the south wall. The analysis of the pressure and temperature differentials across the north and south wall suggested that the difference in drying rates of the north and the south walls was due mainly to the solar effect and not the wind (pressure) effect.

Ranking of the Panels in terms of their Drying Rates

The analysis of the Atlantic Canada Test Hut Moisture Data led to the following ranking of the panels in terms of their drying rates:

Panel 4 - fiberglass sheathing, furred siding
Panel 3 - fiberglass sheathing, no furring
Panel 8 - polystyrene insulation, sheathing paper, no furring
Panel 2 - waferboard sheathing, furred siding
Panel 1 - waferboard sheathing, no furring
Panel 5 - polystyrene sheathing, no furring
Panel 6 - polystyrene sheathing, furred siding
Panel 7 - wet sprayed cellulose insulation

Panel 4 had the fastest average drying time and panel 7 had the slowest average drying time.

WALLDRY COMPUTER PROGRAM [3]

The WALLDRY program is a computer program that dynamically simulates the processes of drying and wetting of wood frame walls over a pre-selected period of time. WALLDRY takes into account the effects of:

- The material properties of the wall assembly (density, specific heat, thermal resistance, moisture diffusion resistance, and if wood based, the hygroscopic properties of wood).

114   WATER IN EXTERIOR BUILDING WALLS

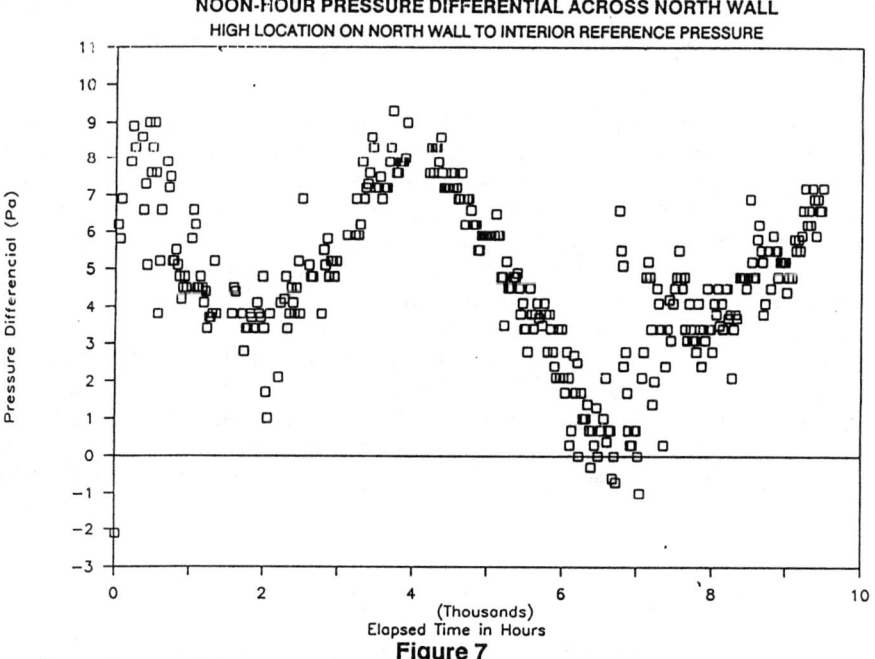

Figure 7

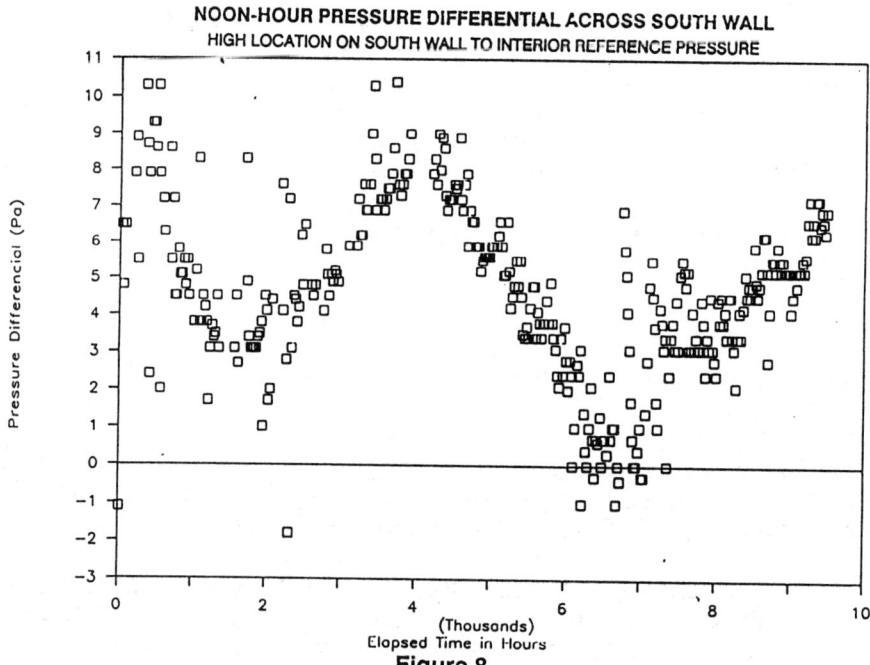

Figure 8

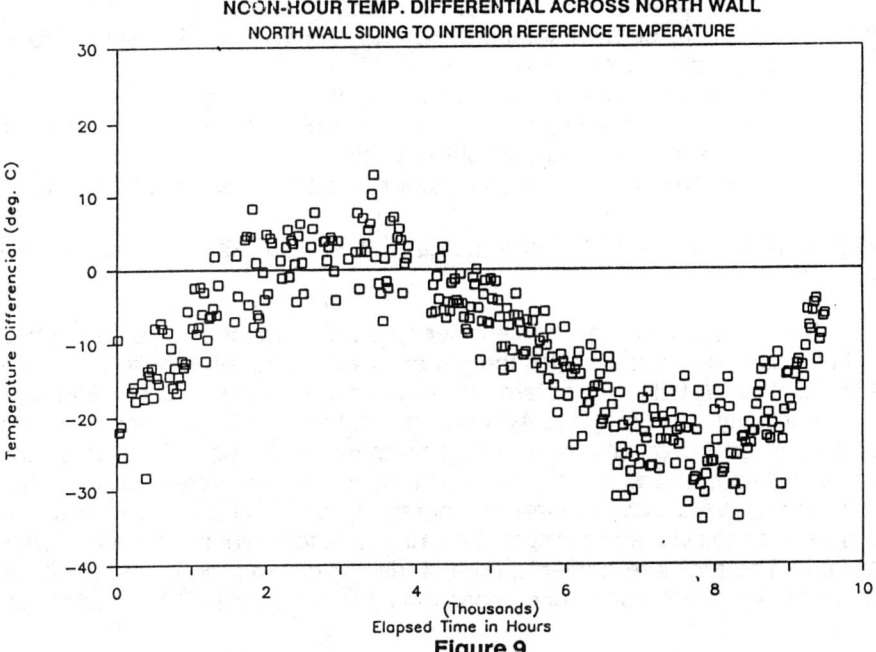

Figure 9

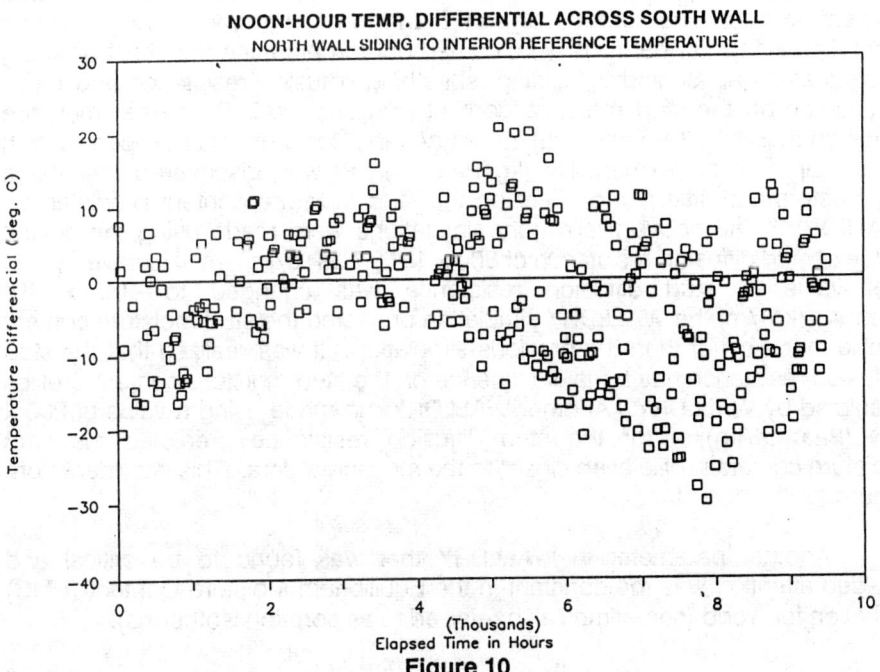

Figure 10

- A full year of hourly weather data including wind speed and direction, outdoor temperature and relative humidity, solar effects and night sky radiation.
- Overall wall dimensions and orientation.
- The configuration of the wall assembly (materials placement, thicknesses, spacing, airtightness, etc.).
- Initial moisture content of the various materials comprising the wall.

Comparison of WALLDRY Predictions with the Atlantic Canada Moisture Test Hut Data.

The Atmospheric Environment Services (AES) weather data for each of the three cities was acquired for the period overlapping the monitoring in the test huts. The WALLDRY program was run using a specific wall configuration in a specific city and using the AES weather data for that city. The time and date for the period of simulation were the same as the period for which the measured data from the Atlantic Canada Test Hut project were available. The results of the WALLDRY simulations, therefore, could be compared with the measured Atlantic Canada Moisture Test Hut data, for validating the WALLDRY program. The moisture content of the studs, which was monitored in all the test panels, formed the basis of comparison against the WALLDRY simulations.

WALLDRY Simulations of the Atlantic Canada Test Hut Panels

A number of preliminary WALLDRY runs were made and the stud moisture content profiles were compared with the measured stud moisture content profiles. The comparison was found to be very poor. This is evident from Figure 11. Further runs were carried out to determine the effect of siding airtightness, gap behind the siding, sheathing diffusion resistance and panel orientation on the stud moisture content profiles. WALLDRY uses moisture diffusion resistivity $((Pa.m^2.s/kg)/m)$ as an input for a material property which is a reciprocal of permeability $(kg/Pa.s.m^2)$. It was discovered that these parameters had little or no effect on the stud moisture content predicted by WALLDRY. The above preliminary simulations were made using the default value of stud diffusion resistance of 800×10^9 $(Pa.m^2.s/kg)/m$. However, when this value of stud diffusion resistance was changed to 400×10^9 $(Pa.m^2.s/kg)/m$, the WALLDRY simulation predicted the stud moisture content profile much better than the previous simulation. It was realized that the stud diffusion resistance has a major influence on the stud moisture content profiles predicted by WALLDRY. Another WALLDRY run made using a value of 200×10^9 $(Pa.m^2.s/kg)/m$ for the stud diffusion resistance predicted the stud moisture content profile even closer to the measured data. This is evident from Figure 12.

Another parameter in WALLDRY that was found to be critical and needed attention was the constant in the Equilibrium Moisture Content (EMC) equation for wood (sometimes also referred to as sorption isotherms).

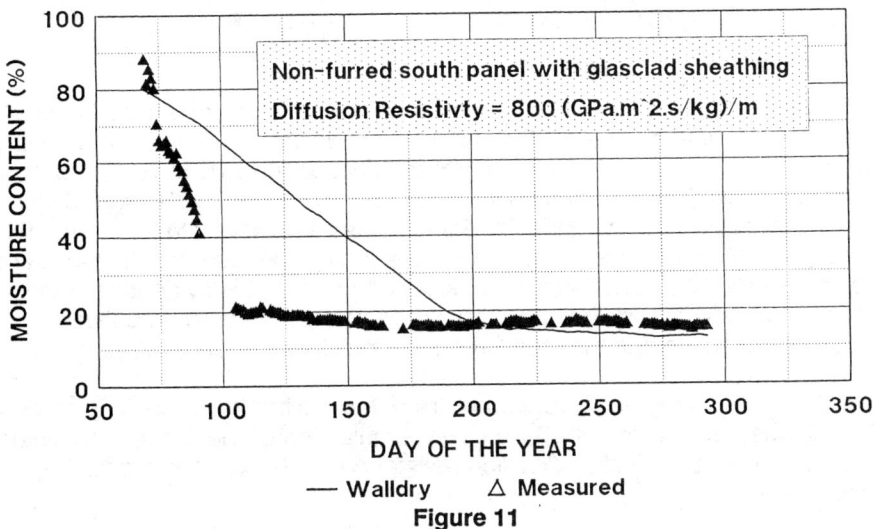

Figure 11

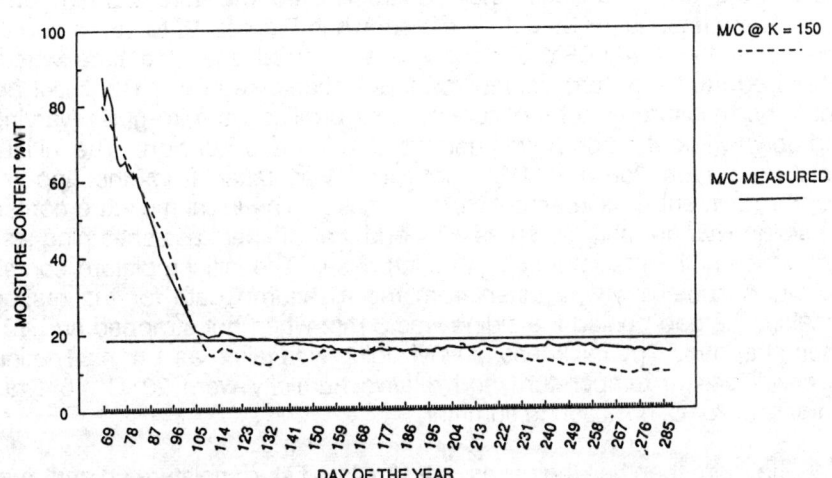

Figure 12

The EMC equation appears in the WALLDRY program in the following form:

$$RH = (MC/6)*K/(EXP(10/6)) \quad \text{for } MC \leq 6$$
$$RH = K/(EXP(10/MC)) \quad \text{for } MC > 6$$
where K = 150.

This relationship was developed by using published data for dry wood. In these two equations, the relative humidity is related to the moisture content of wood. "K = 150" is the parameter whose effect on the stud moisture content profiles was studied by varying it. The test hut panel with high density fiberglass sheathing (panel 3) in St. Johns, was simulated with K = 150, 140 and 120 respectively. The effect of this variation on the stud moisture content predictions of WALLDRY is shown in Figures 12 to 14. It is evident from these runs that lowering K from 150 down to 120 reduced the deviation between the WALLDRY prediction and the measured stud moisture content profiles.

Based on the results obtained from the calibration modeling carried above, it was decided that the WALLDRY simulations of the remaining panels would be made using a stud diffusion resistance of 200×10^9 (Pa.m^2.s/kg)/m and K = 120.

After making the adjustments to the WALLDRY model (constant "K" in the EMC equation) and material properties database, the following panels were simulated:

a) south panels 1 - 4 in each of the three test huts
b) north panels 1 and 2 in each of the three test huts.

The WALLDRY predicted stud moisture content profiles for the above cases were compared with the measured stud moisture content profiles. These comparisons for St. John's are shown in Figures 15 to 18. It should be noted that the WALLDRY simulation was started after the time when the measured stud moisture content profiles became regular. The initial period when the measured stud moisture content profiles were irregular (varying up and down without a trend) was not included in the simulation. The initial stud moisture content for WALLDRY simulation was taken from the upper stud moisture content measurements in each case. The initial moisture content of the siding was normally taken as 30% and that of fiberglass sheathing also as 30% which is the maximum limit in each case. The initial moisture content of the wood sheathing was taken from the measured data for the respective panels. The gap behind the siding was 3 mm when not strapped and 19 mm when strapped. The thickness of vinyl siding was taken as 1 mm. The indoor air conditions of temperature and relative humidity were 20° C, 40% during winter and 25° C, 60% during summer.

In more than half the cases, the simulated stud moisture content profiles were found to agree reasonably well with the measured data. The agreement was particularly good for the huts in St. John's and Halifax. WALLDRY best simulated the panels with fiberglass sheathing (panels 3 and 4). Fiberglass has

JHINGER ET AL. ON COMPARISON OF A WALL MOISTURE MODEL 119

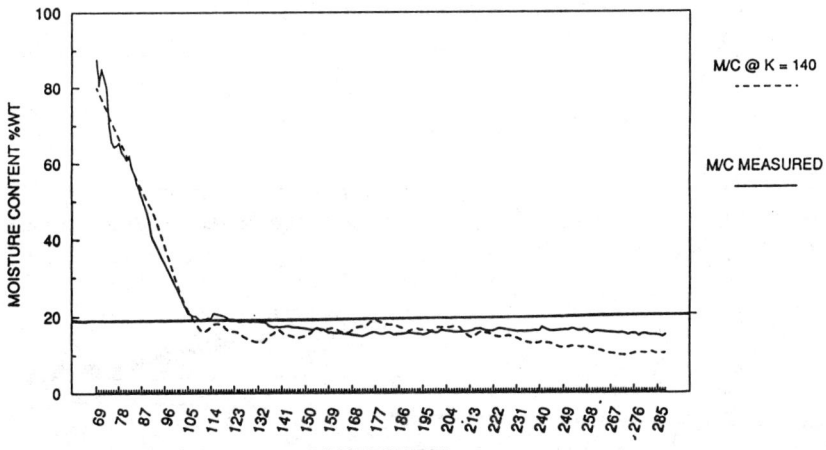

Figure 13

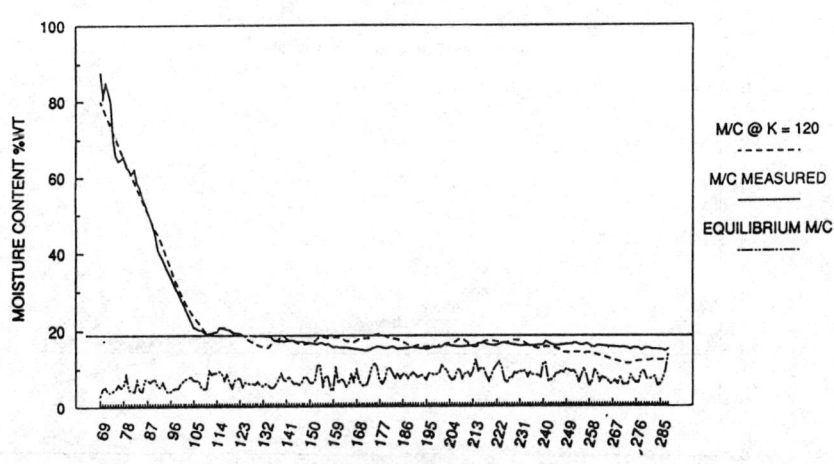

Figure 14

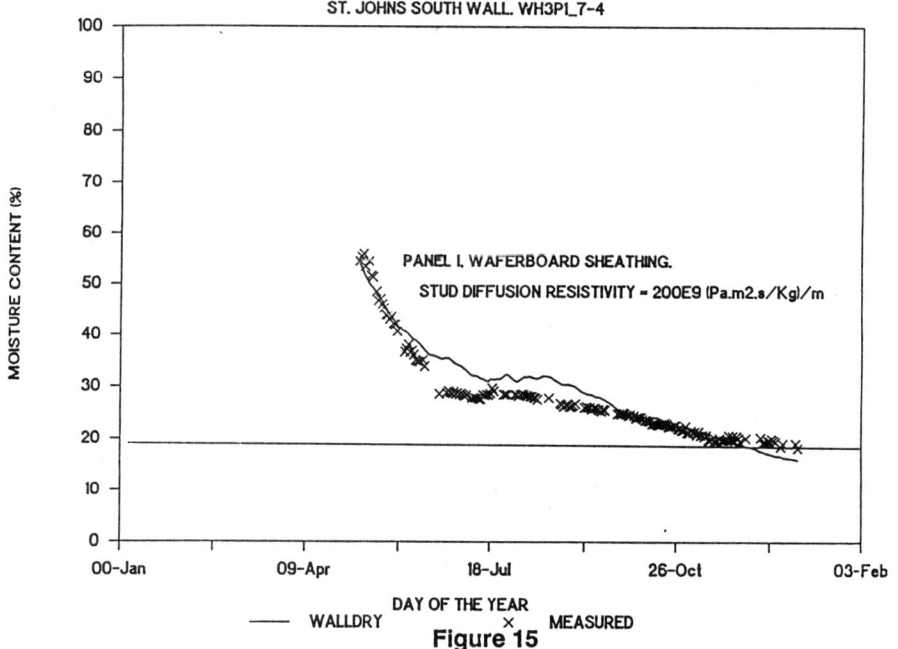

Figure 15

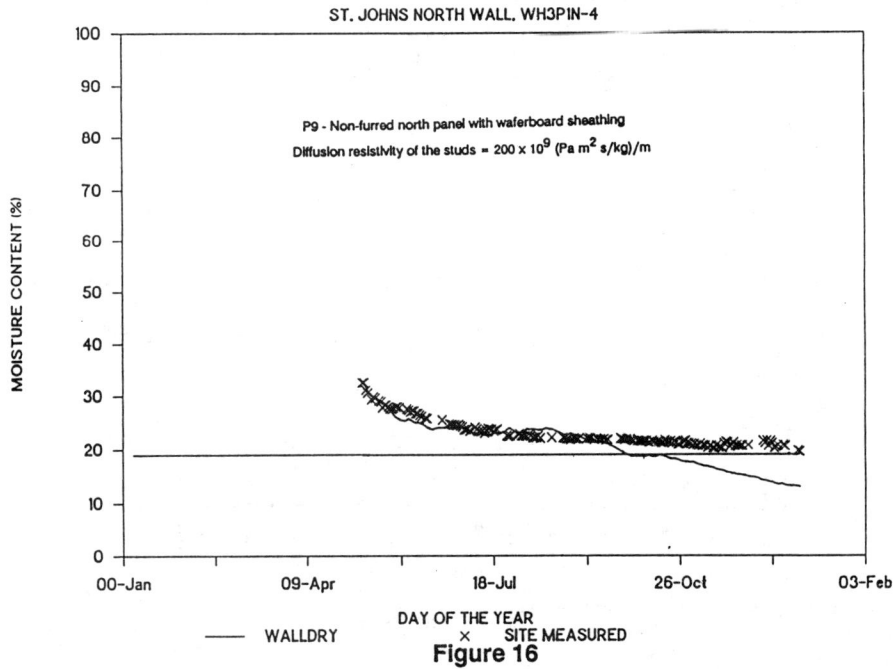

Figure 16

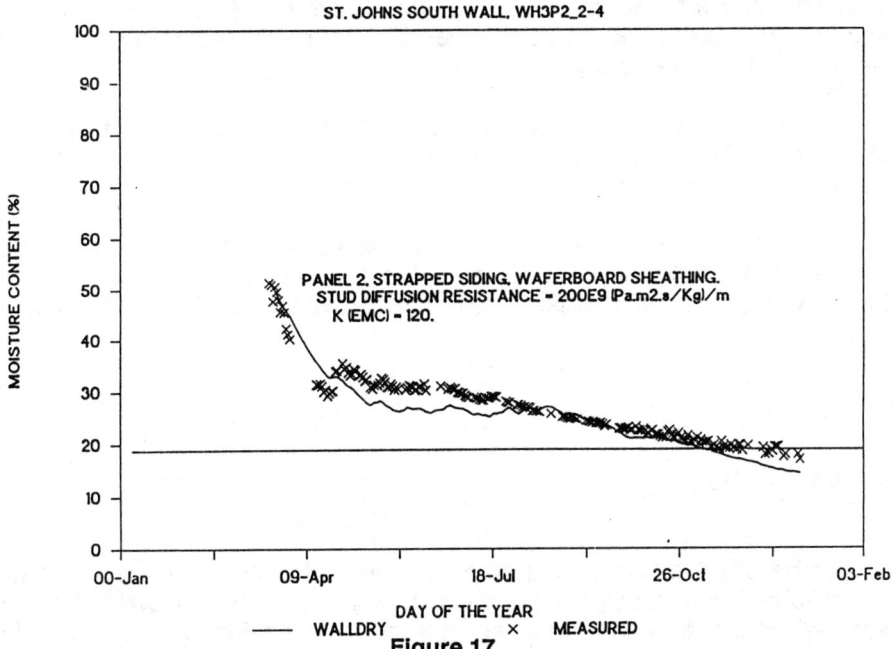

Figure 17

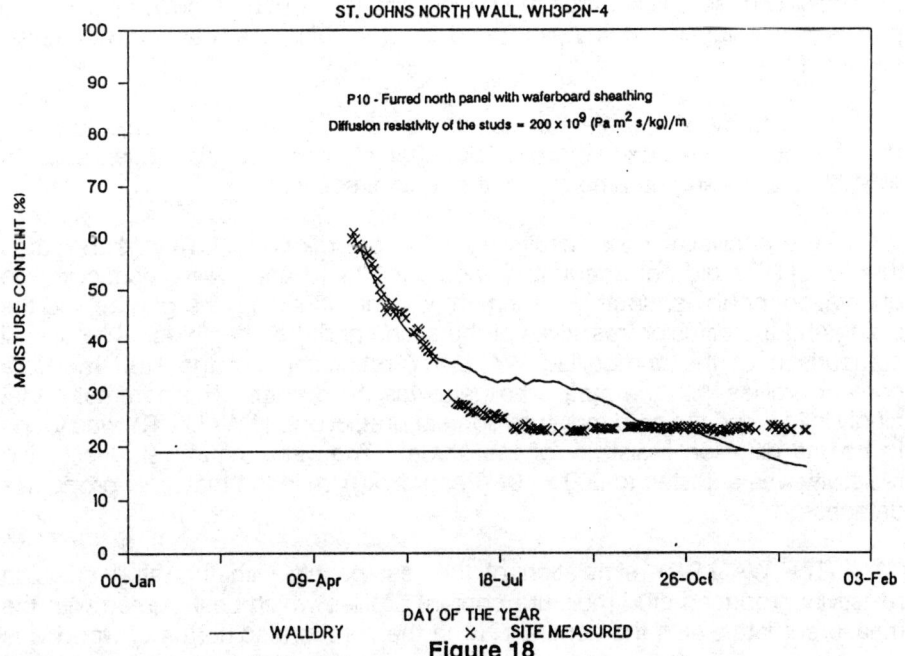

Figure 18

a very low diffusion resistance and therefore makes the moisture transport phenomenon simple as compared with the waferboard sheathing (in panels 1 and 2). WALLDRY needs to be modified in order to better simulate the panels with high diffusion resistivity.

The model showed absolutely no difference in the drying rates of north and south walls. This suggested that the solar model in the WALLDRY program was inadequate and needed improvement.

The WALLDRY model was found to be highly sensitive to any changes in the Equivalent Moisture Content equation. It is therefore speculated that the EMC equation may be modified in order to improve the accuracy of the model.

CONCLUSIONS

The effect of the furring strips on the drying of wall panels with low permeabilities was not significant when vinyl siding is used. However, panels with high permeability showed a decrease in the drying time when furring strips were used behind vinyl siding.

The permeability and the airtightness of the sheathing was a decisive factor in determining the drying rates of the panels. The panels with higher permeability dried quicker than the panels with lower permeability. Similarly the panels with higher airtightness dried slower than the panels with lower airtightness.

The study of the effect of panel orientation on the drying rate revealed that the south panels dried more quicker than the north panels. Solar radiation was the determining parameter, not the wind direction.

The analysis of the results from the preliminary simulations indicated that WALLDRY did not adequately model the test panels with respect to the orientation (north vs. south), the airtightness of the siding, the gap behind the siding and the diffusion resistivity of the siding and the sheathing. The overall comparison of the unmodified WALLDRY predictions of the stud moisture content profiles with the measured data was quite poor. The parameter that highly influenced the stud moisture content predictions of WALLDRY was found to be the diffusion resistivity of the studs. The value of the stud diffusion resistivity was adjusted to 200×10^9 (Pa.m^2.s/kg)/m in the materials properties database.

The WALLDRY simulation of the test panels with this stud diffusion resistivity produced stud moisture content profiles which best agreed with the measured data. Also the constant "K" in the sorption isotherms of wood was adjusted to 120 (from 150). All further WALLDRY runs were carried out with these changes incorporated.

Using these modified property values, the WALLDRY simulations of more than half the panels predicted stud moisture content profiles that agreed reasonably well with the measured data. The agreement was particularly good for the test panels containing fiberglass sheathing (low diffusion resistivity), where the transfer of moisture is basically a diffusion phenomenon. The test panels with high diffusion resistivity showed poor results probably because in this case condensation on the interior of the sheathing is an important element in the process, and this phenomenon seems to be inadequately modeled in WALLDRY. The moisture diffusion resistivity of wood should be a function of the wood moisture content.

The solar model responded poorly with respect to the wall orientation. WALLDRY under-predicted the wall surface temperature (during the day) and over-predicted the nightsky radiation. The removal of heat from the wall surface (exterior) via convection was not modelled adequately by WALLDRY. The convective heat transfer co-efficient during day should be made to vary with the direction and speed of wind incident on the wall surface.

The overall validation of WALLDRY suggested that it can simulate the drying of walls reasonably well if suitable material properties of wall components are utilized. The program is being modified by making the necessary improvements to the solar and moisture models.

APPLICATIONS OF WALLDRY COMPUTER PROGRAM

WALLDRY computer program is intended to be used as a research tool and should be of interest to researchers and engineers in the building science community, the construction industry and the building materials manufacturing sector. In view of its research nature, the results from WALLDRY should not form the only basis for design or installation decisions, but should be supplemented by hard research data and practical experience.

REFERENCES

[1] Oboe Engineering Ltd. — Manual for Atlantic Canada Hut Project, March 6, 1989.

[2] Canadian Mortgage and Housing Corporation — CMHC/CHBA Task force on moisture problems in Atlantic Canada.

[3] By SCANDA consultants and RWDI for CMHC — Wall Drying Simulator (WALLDRY) Version 1.0, December 1988.

Robert J. Kudder and Kurt R. Hoigard[1]

VAPOR CONTROL AND PSYCHROMETRIC MONITORING IN EXTERIOR WALLS

REFERENCE: Kudder, R. J., and Hoigard, K. R., "Vapor Control and Psychrometric Monitoring in Curtain Walls", Water in Exterior Building Walls: Problems and Solutions, ASTM STP 1107, Thomas A. Schwartz, editor, American Society for Testing and Materials, Philadelphia, 1991.

ABSTRACT: Curtain walls may contain a cavity between the interior finish wall and the exterior cladding. Condensation can occur in this space as interior moisture enters either through direct air leakage paths or diffusion through construction materials, and can significantly affect the facade life expectancy. Buildings exhibiting facade degradation due to cavity moisture generally require detailed evaluation in order to determine the source, and to develop an appropriate repair plan. One tool available to the investigator is field testing. This paper will address typical wall behaviors and the practical aspects of field instrumentation to acquire data for determination of the presence or absence of cavity condensation, and a time history of its occurrence.

KEYWORDS: condensation, monitoring, psychrometry, relative humidity, thermocouples, transducers, vapor

There are many incentives to introduce and maintain controlled levels of interior humidity. Mechanical systems that supply water vapor for humidification are common features of contemporary buildings. The desired level of interior relative humidity varies with the building occupancy. In hospitals, humidification was originally intended to control sparks from static electricity near ether. It was found that humidification also improved the comfort level of patients, resulting in its continued use. Electronic equipment, such as used in hospitals and modern offices, and machines which handle paper, perform more reliably and suffer less from the effects of static electricity in buildings with controlled humidification. Wooden musical instruments, furniture, art objects and archive materials held by libraries and museums also benefit from an environment with a constant and controlled relative humidity. Residential occupancies benefit from improved comfort and lower winter operating temperature with controlled interior relative humidity levels.

In cold climates, the difference between the warm humidified interior and the cold dry exterior creates partial vapor pressure differences which drive the interior moisture outward. The building envelope must confine the interior water vapor and maintain it above the dew point temperature. If the envelope fails to do this, condensation will occur. Facades have been severely damaged by the accumulation of water and ice due to condensa-

---

[1]Dr. Kudder is a Principal, and Mr. Hoigard is Manager, Testing Services, with Raths, Raths & Johnson, Inc., 835 Midway Drive, Willowbrook, IL 60521.

tion. The expansive pressure from ice can fracture or move wall components. Water accumulation in wall materials can cause deterioration, change insulating properties, support bacteria growth, and accelerate corrosion. The useful life of the wall will be shortened.

Approximate design procedures are available to predict the temperature gradient and the partial vapor pressure gradient through a wall under given interior and exterior conditions. The procedures are based on a simple linear conductive heat flow model and a linear diffusion model, respectively [1][2][3]. If the thermodynamic behavior of a wall is reasonably consistent with these assumed models, the design procedures have proven useful and successful.

Unacceptable thermodynamic performance and damage can result from two causes. First, thermodynamic behavior might have been ignored in the wall design. This is often predictable by simple design procedures and therefore could have been avoided. Analyzing such a wall is relatively straightforward. Second, even if design procedures were appropriate, poor details and as-built construction can result in a wall which does not actually behave in a manner reasonably consistent with simple linear design models [4]. Analyzing such a wall can be immensely complex and unreliable. Acceptable and predictable thermodynamic behavior achieved through straightforward design procedures, good constructible details, and careful construction, is a characteristic of a successful wall. Current design trends are leading to increased predictability and reliability by including distinct wall components to control the movements of heat, vapor and air. Conversely, thermodynamic behavior which differs significantly from linear conductive heat transfer and vapor diffusion models can be a characteristic of a poorly performing wall. Identifying and quantifying the differences between design models and actual behavior is useful in understanding the cause of poor performance and in determining repair requirements.

This paper discusses the behavior of walls with acceptable and unacceptable vapor control, and presents the implementation of field instrumentation to monitor the thermodynamic behavior of curtain walls. The interpretation of the data which indicate specific aspects of the wall behavior is also discussed.

WALL BEHAVIOR

Table 1 summarizes the features and responses of walls which function properly, and walls which exhibit performance problems in cold climates. A wall which functions properly meets the expectations of the design, limits heat loss and air infiltration/exfiltration, and controls vapor movement. Condensation is likely to occur only under exceptional conditions, and is not likely to cause damage or deterioration.

Problem walls resulting in uncontrolled and damaging condensation can have either too little or too much vapor control. Inadequate vapor control can result from diffusion of vapor through the wall materials, allowing the vapor to reach zones where the partial vapor pressure approaches the saturation vapor pressure. Materials with permeance ratings which are too high for the driving partial vapor pressure, or the omission or penetration of a vapor retarder, result in excessive diffusion and consequent condensation. Excessive diffusion has also been the cause of damage in recently renovated older buildings, where insulation has been added or upgraded without consideration of vapor control. The added insulation lowers the temperature of outer wall components, and therefore lowers the saturation vapor pressure and increases the likelihood of condensation in the bricks.

TABLE 1 -- Comparison of Wall Response Behaviors

WALL FEATURES	THERMAL RESPONSE	VAPOR RESPONSE	AIR RESPONSE
Properly detailed and well constructed: 1. Insulated 2. Incorporates vapor control 3. Incorporates air control	1. Insulation functional and effective 2. Temperature gradient approximated by conductive heat model	1. Vapor retarder intercepts flow to exterior 2. Partial vapor pressures kept below saturation vapor pressures 3. Condensation is controlled 4. Wall materials in equilibrium and not repeatedly saturated 5. Partial vapor pressure gradient approximated by diffusion model	1. Limited infiltration/exfiltration 2. Limited air exchange from interior to cavity 3. Limited air transport of interior vapor 4. Limited air paths for unobstructed diffusion of vapor

Inadequate vapor control:
1. Insulated
2. Lacks vapor control
  a. Excessive diffusion
  b. Air paths
3. Lacks air control
	1. Insulation becomes wet or saturated
	2. Insulation bypassed by air paths and convection currents
	3. Partially effective insulation lowers temperature of wall materials below dew point
	4. Overall evaporative cooling
	5. Unpredictable effects from convection
	6. Temperature gradient not approximated by conductive heat model
		1. Excessive diffusion due to breaches in the retarder or high perm ratings
		2. Air paths for transport or unobstructed diffusion of interior humidity into the cavity
		3. Partial vapor pressure near the saturation pressure
		4. Condensation within the wall
		5. Freeze-thaw and moisture related damage to materials
		6. Vapor pressure gradient not approximated by diffusion model
			1. Excessive air infiltration and exfiltration
			2. Difficulty in controlling interior environment due to drafts and loss of humidification
			3. Excessive energy costs
			4. Odors from mildew and mold on wall materials possible

Excessive vapor control:
1. Insulation may be incorporated but mislocated
2. Vapor movement inhibited
  a. Barrier mislocated or incompatible with insulation location
  c. Barrier unrecognized or ignored
3. Incorporates air control
	1. Temperature gradient may be approximated by conduction model unless insulation layers become wet from condensation
	2. Temperature gradient may allow condensation
	3. Condensation may occur in exterior layers, where temperature is below freezing
		1. Vapor migrates but is prevented from discharging to the exterior
		2. Condensation and accumulation of moisture occurs
		3. Vapor pressure gradient may actually be approximated by diffusion model, and therefore be predictable
			1. Evaporation of condensate prevented
			2. Venting to transport vapor is inhibited
			3. Odors from mildew and mold on wall materials possible

A potentially more serious inadequacy in vapor control is an air path which permits communication between the conditioned interior of a building and the cavity within a wall. Vapor can be transported by moving air, bypassing the vapor retarder and insulation, and reaching zones of the wall where temperatures are not kept above the dew point. Even if air pressure differences do not cause interior air to flow toward the cavity, air paths provide open and unobstructed avenues for diffusion. The mass of water that can move along air paths is usually higher than the mass that can move by diffusion through wall materials, and the resulting condensation problem can be more severe. Air paths result from a failure to seal the interior and isolate it from the cavity. Unsecured drywall, unsealed gaps at the top of masonry backup walls, termination of interior walls above dropped ceilings, unclosed column enclosures, open flutes of ribbed metal decking, open cores of precast concrete plank, unsealed joist bearing pockets, unsealed utility penetrations, and other detailing and construction deficiencies create these air paths.

Condensation damage will also result in a wall that has excessive vapor control. This condition is best described as having too much vapor control in the wrong place. Vapor will condense at a barrier with low permeance if its temperature is not kept above the dew point. The barrier may be an intentional component of the wall which is located incorrectly. The barrier can also be unintended or unrecognized. The most common unrecognized barriers are the glazed faces of bricks, waterproofing and finish coatings with low permeability, and uninsulated metal components. Vapor which reaches these uninsulated barriers by diffusion or air paths, or both, will condense.

Walls with air paths sufficient to cause condensation problems cannot be analyzed using simple linear conductive heat transfer and diffusion models. The effects of mass transport, convection, and differential air pressures must be accounted for. There are numerous parameters for such an analysis, and the computations are complex. To understand the behavior of a wall with air path problems, the writers have found it practical to measure the psychrometric response of the wall.

RESPONSE MEASUREMENTS

Measuring the psychrometric time history of a wall provides information necessary to determine if condensation occurs, the circumstances under which it occurs, the influence of infiltration from the exterior, the effect of air paths and diffusion from the interior, the overall effectiveness of the design, or the success of remedial measures. Transducers embedded in the wall and controlled by a data acquisition system are used to periodically measure temperatures, relative humidities, and differential air pressures.

By carefully selecting the types of sensors used and where they are placed, physical processes underway in the wall can be monitored. The sensors selected must be capable of measuring a parameter with sufficient range, accuracy and sensitivity, and be capable of reporting it in a form compatible with the recording equipment being used. Measurements typically of interest to the curtain wall investigator include temperature, relative humidity, air pressure, and environmental wind speed and direction. Using basic parameters such as these, additional information may be determined by using psychrometric equations [5].

<u>Temperature</u>

Most people are familiar with mercury thermometers. These devices are simple and easy to use, but cannot be read from a remote site. Transducers used for wall monitoring need to have an electrical output proportional to or predictably related to the physical value being measured in order to allow the use of a data acquisition system. Two temperature measuring devices that meet this criteria are thermocouples and resistance temperature detectors (RTD).

Thermocouples produce a small DC output voltage that can be converted to temperature through the use of a calibration curve. The calibration is reasonably linear over a range of temperatures, and is usually built into the thermocouple indicator so that it reads directly in degrees. While there are many types of thermocouples designed for different temperature ranges, the type J thermocouple is well-suited for monitoring curtain walls; its linear range extends well beyond the temperatures commonly encountered [6]. Caution is advised, however, when procuring type J thermocouple indicators. Some manufacturers' equipment will not read temperatures below freezing accurately, even though the thermocouple itself can perform adequately to temperatures of minus 40 degrees C (minus 40 degrees F).

RTDs produce a change in electrical resistance that is correlated to temperature change. When placed in a constant current circuit, the voltage drop across the RTD can be measured and related to temperature. Like thermocouples, the selection of RTDs must be made based on the temperature range expected to be encountered.

Temperature sensors are placed in, on and around the subject wall at locations that will allow a temperature gradient through the wall to be determined. Locations should be selected to provide both air temperatures and material surface temperatures. Typical investigations require the measurement of building interior, exterior and cavity air temperatures along with exterior and interior surface temperatures.

In-place protection of temperature sensors is required in most installations, particularly for those sensors positioned to measure outside air and cavity air temperatures. Type J thermocouples contain soft iron and must be protected against corrosion. A thin coating of lacquer applied after the two thermocouple conductors are twisted together has proven to work well. Outside air temperature sensors must be shielded from direct sunlight to avoid solar heating. Figure 1 illustrates an outdoor air temperature enclosure designed to shield the sensor from sunlight and precipitation while allowing adequate ventilation.

Relative Humidity

Many types of relative humidity measuring devices are available [7]. Some materials, such as animal hair or special plastic filaments, change length when subjected to varying relative humidity. Other materials produce either a change of electrical resistance or a change in capacitance with a change in relative humidity. The latter two are most frequently incorporated in electronic instruments used in psychrometric monitoring programs. Variable capacitance and variable resistance sensors can both be integrated with additional electronic circuitry to produce a transducer with a linearly varying DC voltage output proportional to the measured relative humidity. Figure 2 shows a variable capacitance humidity probe. The plastic jar on the end is a calibration standard which uses a saturated salt solution to produce a known relative humidity.

When selecting a relative humidity probe, several important characteristics must be considered. Transducer accuracy, sensitivity, response time and range of operation should reviewed and checked against the requirements of the investigation. Accuracies of ± 1 to ± 2 percent R.H. are usually sufficient. It is common for the accuracy of a remote reading

FIG. 1 -- Enclosure sheltering a thermocouple used to measure outside air temperature.

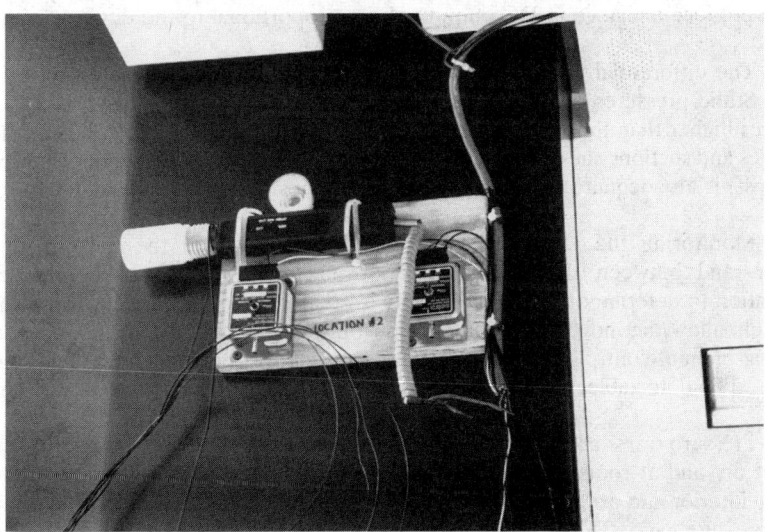

FIG. 2 -- Capacitance type humidity transducer shown with a calibration bottle on the sensor. Also shown are two differential pressure sensors.

humidity transducer to be less at or near 100 percent relative humidity than at mid-range humidities. If sub-freezing temperatures are anticipated, an instrument capable of accurate measurements of relative humidity over ice as well as over water is needed. When condensation is expected, the ability of the instrument to recover from wetting becomes important. Long term monitoring requires instruments that exhibit little or no hysteresis effects.

Placement of humidity probes should be such that the relative humidity gradient through the wall being studied can be determined. Typically, this requires the transducers to be installed so that outside air, inside air and wall cavity air can be sampled. Air temperature should be measured at each humidity probe to enable dew point and specific humidity calculations to be performed.

The electronic circuitry which converts resistance or capacitance change to a DC output voltage must be kept dry and within its operating temperature range. This is accomplished by using instruments with remote probe tips connected to a separate housing for the electronic circuits by a lead wire. By keeping the main housing of the unit in an interior building space while placing only the probe tip in the measurement zone, the majority of the components can be protected. Also, the probe tip itself should be kept free of precipitation and the contaminants often associated with cavity condensate. The use of a drip cap will suffice in most cases.

Air Pressure

As with temperature and relative humidity measurement, there are many types of pressure measurement devices. Absolute pressure transducers are expensive and not well suited for monitoring wall behavior. To determine the pressure drop across a wall using absolute pressure transducers requires subtracting very small numbers, which can lead to large errors. The most appropriate types for monitoring wall behavior are those that produce a change in resistance or capacitance due to a pressure differential across a diaphragm. These devices produce a change in DC output voltage proportional to the differential pressure.

The differential pressures typically encountered in building walls are actually quite small. Static pressures generated by HVAC equipment rarely exceed 0.096 kPa (2 psf); pressures higher than this would make it too difficult for occupants to open and close doors. Pressures and suctions caused by winds can be far larger, but experience has shown ± 1.2 kPa (± 25 psf) is an adequate measurement range for most midrise buildings.

Monitoring the relative pressure differences between the building interior and exterior, and between the building interior and wall cavity usually provides enough information to determine whether the cavity is isolated from the interior or if air paths exist that could allow the movement of interior air outward into the cavity. Supplementary tests involving manual control of building HVAC fans can, if building static pressures are high enough to be detectable, be used to determine if the cavity communicates with the interior.

Pressure transducers are generally placed within the building interior, where they can be kept dry and at room temperature. To measure the pressure differential between the building interior and exterior, a reference tube must be routed to the exterior and connected to the pressure gage. This tube should be no smaller than 3.063 mm (1/8 in.) in diameter and no longer than 3 meters (10 ft.) to minimize the damping effects of the column of air in the tube. The outside end of the tube is best placed flush with the building exterior finish. It should be routed with an upward slope of at least 45 degrees so that rain water which enters it can drain and not become airlocked in the tube.

Weather Station

Exterior air temperature and relative humidity should always be included in measurements made for a wall investigation. In addition to these, wind speed and direction are also useful. Together, these four measurements comprise a weather station. Data retrieved from an on-site weather station can be compared to data from the National Weather Service (NWS) [8] for the nearest monitoring station. There is a time delay of many months in the publication of NWS data, and it may be necessary to visit the recording station and transcribe the data manually. This comparison permits not only a check of the instrumentation system, but an extrapolation of results for periods not monitored. Figure 3 shows a combination wind speed anemometer and wind direction indicator mounted on a tripod on the roof of a building.

An important consideration when choosing anemometers to monitor wind speed is the form of their output. Devices with an AC output voltage proportional to wind speed are less expensive than DC output units, but are more difficult to read. The AC signal must be sampled at a fast enough rate and for a long enough period of time to allow an RMS calculation to be performed. This method of sampling the output is preferable to the use of full-wave rectifier circuitry for conversion to a DC signal since most rectifiers are susceptible to signal noise spikes.

FIG. 3 -- Wind speed anemometer and direction indicator installed on the roof of a test building.

INSTRUMENTATION PROGRAM

Once a reliable set of instruments has been selected, a psychrometric monitoring program can be implemented. A comprehensive plan of instrumented locations, data recording hardware, data collection rates and intervals, and the duration of the monitoring period is needed to insure a useful end result.

<u>Instrumented Locations</u>

Economic concerns usually limit the number of instrumented locations on a project to three or four. Access to the monitoring location on the interior and the exterior, and routing of lead wires through the building, are major cost considerations. The effectiveness of an instrumentation program can be greatly enhanced if care is taken in the selection of these areas. A review of the project documents and as-built conditions can help find potentially vulnerable locations. Areas displaying visible symptoms of poor performance, such as deterioration, water accumulation, staining, ice formation and efflorescence, can be determined by condition surveys. Infrared thermography can be used to study concealed conditions. Figure 4 is an infrared image of a building with unexplained areas of high heat loss on the exterior masonry. Subsequent investigative openings in these areas uncovered large openings in the vapor barrier at structural steel connections. An area such as this warrants monitoring.

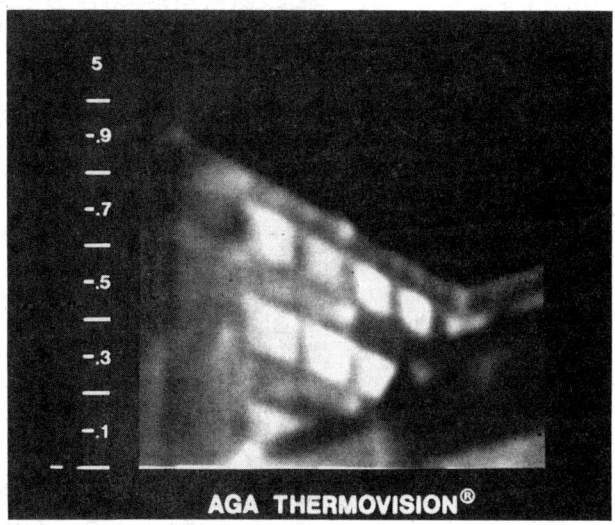

FIG. 4 -- Infrared thermograph used in choosing monitoring locations.

In addition to problem areas, other locations should also be considered for instrumentation. Areas of the building that appear to be performing properly and show no signs of deterioration provide a good baseline for comparison with data from problem areas. Likewise, monitoring areas of trial repairs can demonstrate the degree of their effectiveness.

Data Recording Equipment

Instruments installed for wall monitoring purposes must have their output sampled and stored on a periodic basis in order to generate a time history of wall response. These tasks are typically handled either by PC based data acquisition systems or by traditional data loggers. Modern PC based systems usually consist of hardware for channel multiplexing, DC voltage measurement and data conversion to digital form. Low level signal amplifiers and thermocouple reference junctions may also be incorporated in the system. A computer program operating on the host computer is in control of data sampling and recording; the flexibility of the system is generally limited only by the scope of the program. Advantages of a computer controlled system include enhanced control of data scans and triggering, real time data reduction and plotting, remote control and data retrieval via modem, versatile report generation, periodic retrieval of backup data via floppy disk and unattended restart of the system in the event of a power outage.

An alternative to PC based systems is the data logger. These systems are self-contained units capable of performing all of the tasks associated with reading and storing data. Control is usually accomplished by selecting modes of operation on a control panel, and data may be stored on magnetic tape or disk. An obvious advantage to this type of system is its simplicity, but it may also lack flexibility.

Data Sampling

A program of data read scans spaced at one to three hour intervals generally produces a time record with sufficient detail for analysis purposes. If data reads can be coordinated with the read times at the nearest National Weather Service station, usually the local airport, the additional benefit of a direct comparison between test data and official weather data can be obtained.

In most buildings, some amount of electromagnetic interference (EMI) or radio frequency interference (RFI) can be expected. Data averaging should be employed to reduce sampling errors from these sources. Multiple data samples should be taken over a short period of time and averaged to obtain a mean value for that interval reading. Other strategies for reducing the effects of EMI and RFI include keeping sensor cables as short as possible (preferably under 150 ft.) and using shielded twisted-pair instrument wire. In cases where instrument lead wires must be very long, or noise levels are high enough to adversely affect the data, remote digitizing equipment at the transducer location, combined with digital data transmission, may prove to be the only solution.

System Location

Although it is advantageous to keep instrument lead lengths to a minimum, the central location chosen for the data recording system needs to address other concerns as well. An indoor location with a stable temperature is the best choice. The system location must be safe from theft and vandalism, and also safe from curious bystanders. Reliable line power, grounded outlets and surge protected circuits are also important in minimizing data loss and equipment damage. In addition, proper ventilation is necessary to prevent equipment damaging heat build-up. Poor ventilation can also cause a large change in the temperature of the thermocouple reference junction. For this reason, small unventilated enclosures which are sealed to make them dustproof or rainproof can cause incorrect thermocouple readings unless the reference junction temperature is monitored.

Duration of the Monitoring Program

Psychrometric monitoring programs are most effective when conducted during the winter months, when condensation in susceptible cavity walls may occur. Under winter conditions in a wall in which condensation will occur, the psychrometric properties of moist air result in a change of state of the moisture in the air. These psychrometric events are the key to understanding the behavior of the wall. Therefore, an instrumentation program of this type is most successful when initiated prior to weather conditions likely to cause condensation, and when monitoring is continued long enough to capture the event of condensation in the data. If initiated at the right time, monitoring an instrumented building for a period of four to six weeks has been found to provide adequate data to evaluate the behavior of the building.

Once condensation occurs, and a large quantity of moisture accumulates in the wall materials, the response of the wall may become very complex. Interpretation of the data is complicated by the evaporation of moisture from the wall materials.

INTERPRETING THE TIME HISTORY DATA

Plots of the recorded data can be generated directly by the software which controls the data acquisition system, or the data can be imported into spreadsheet or business graphics programs for plotting. Interpretation of representative data plots to ascertain various aspects of wall behavior is discussed below. The data were recorded during monitoring programs of real buildings.

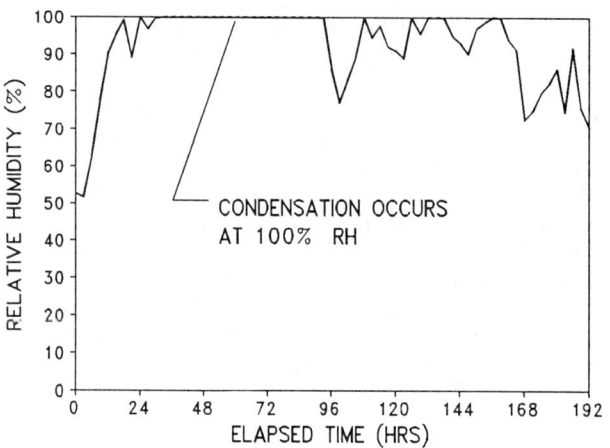

FIG. 5 -- The relative humidity of cavity air, showing several occurrences of condensation.

Condensation occurs when surface temperatures fall below the dewpoint or when the relative humidity of the cavity air reaches 100 percent. The latter has occurred several times in the record shown in Figure 5. Condensation in the cavity air continued for almost 60 hours during the first occurrence. The inside surface temperature of exterior cladding can, at times, be lower than the cavity air temperature, causing condensation to occur on this surface even when the relative humidity of the cavity air is less than 100 percent.

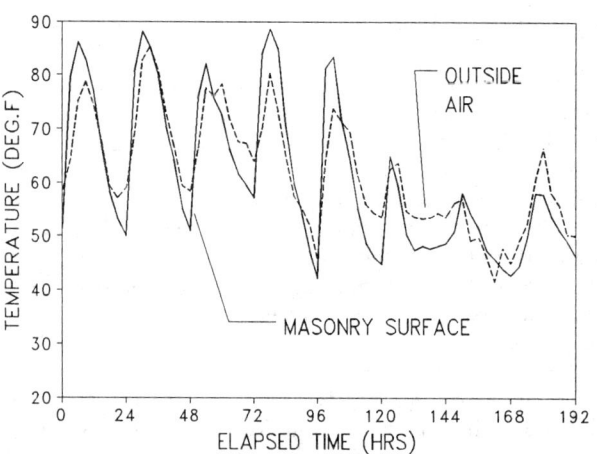

FIG. 6 -- Comparison of outside air temperature and exterior surface temperature for a brick cavity wall.

One of the consequences of condensation in a brick cavity wall is the possible saturation of the brick, and freeze-thaw damage. It can also lower the surface temperature of the building by evaporative cooling. Figure 6 show a comparison between the outside air temperature and the exterior surface temperature of a brick cavity wall which was known to have experienced condensation before and during the time interval plotted. The outside air temperature trace exhibits the expected oscillation between a daily high and low. During the daytime, the masonry surface temperature is higher than the air temperature, as would be expected because of solar heating. However, at night when the air temperatures are low, the masonry surface temperature is even lower. This would not happen if the temperature gradient through the wall could be approximated by a conductive heat transfer model. Radiant cooling may account for a portion of this difference, but it is judged to be caused mainly by evaporative cooling.

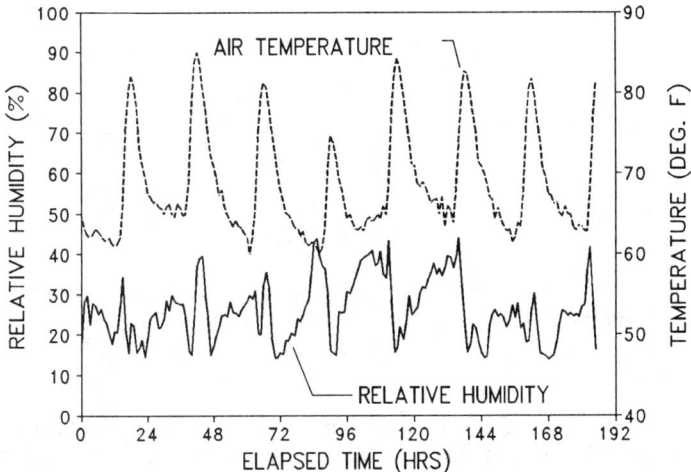

FIG. 7 -- Comparison of cavity relative humidity and cavity air temperature.

Aspects of wall behavior can be interpreted from the relationship between cavity relative humidity and cavity air temperature. For the first portion of the record shown in Figure 7, the cavity relative humidity is approximately in phase with the cavity air temperature. When the temperature increases, the relative humidity increases. This is not how an ideal volume of air containing a constant mass of water vapor would respond to changes in temperature. For the ideal volume of air, the temperature and relative humidity would be out of phase, and lowering the temperature would result in an increase in the relative humidity. This behavior is exhibited in the last portion of the record. When an increase in temperature causes the relative humidity to increase, moisture already trapped in the wall materials is being released into the cavity air. Some of the released moisture leaves the cavity, and some appears to return to the wall materials when the temperature decreases and is available for the next cycle. The material property governing this behavior is sorption. Additional research to understand this property is essential to understanding wall behavior [9].

The air pressure in a cavity will be affected by both exterior and interior conditions, unless the cavity isolation is perfect. Isolation of the cavity from the interior can be studied with data from differential pressure transducers while the fans of the building HVAC system are turned off and on over a short period of time. The rate at which equilibrium is reached, after a change in interior pressure, is affected by the degree of cavity isolation. The data plotted in Figure 8 were obtained from a project in which transducers measured the differential pressure between the cavity and the interior of the building (P_{ci}), and between the exterior and the interior of the building (P_{ei}). The ratio of these two measurements, $P_{ci} \div P_{ei}$ indicates the portion of the pressure difference between the cavity and the interior

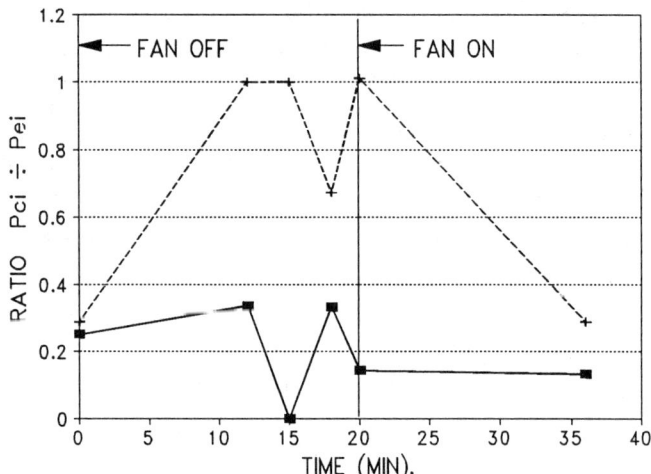

FIG. 8 -- The ratio of differential pressure between the cavity and the interior to the differential pressure of the exterior to the interior, $P_{ci} \div P_{ei}$, plotted for two test locations.

which can be attributed to the exterior environment. Studying this ratio rather than the individual differential pressure measurements normalizes the effect of the exterior environment. If this ratio is approximately one, the air pressure in the cavity is essentially responding to the exterior environmental conditions. If the ratio is approximately zero, the air pressure in the cavity is essentially responding to interior conditions. The closer the ratio is to zero, and the more rapidly the ratio approaches zero after the building fans are turned off, the greater the communication by air paths between the interior and the cavity. In Figure 8, the response at the location shown by the dashed line indicates better isolation between the cavity and the interior than at the location shown by the solid line.

CONCLUSIONS

In cold climates, condensation in cavity walls can cause deterioration and damage. If the wall behavior is reasonably consistent with linear heat transfer and diffusion models, condensation can be predicted using traditional calculation methods. If the wall has air paths through which the cavity can communicate with the interior, traditional calculation methods do not apply, and applicable calculation methods can be so complex as to be impractical. For problem walls with air paths, it is more useful to instrument the wall and measure its thermodynamic response.

A successful monitoring program requires careful planning and a practical consideration of the following:

1. Selection of appropriate transducers for each parameter to be measured.
2. Selection of monitoring locations which include both acceptable and unacceptable apparent performance.

3. The number of instruments at each monitoring location sufficient to determine the temperature and partial vapor pressure gradients through the wall.
4. The location and configuration of the data acquisition system, including scanning rates.
5. Initiation of the program so that the event of condensation can be captured in the data.
6. Continuing the monitoring program long enough to acquire data on the behavior of the wall after condensation occurs.

Evaluating the impact of condensation and designing wall repairs to limit condensation requires an understanding of the thermodynamic behavior of the wall. This behavior can be interpreted from data obtained with a psychrometric monitoring program.

REFERENCES

[1] ASHRAE Handbook of Fundamentals, American Society of Heating, Refrigerating and Air-Conditioning Engineers, Atlanta, 1989, Chapters 20, 21 and 22.

[2] "Moisture Control in Brick and Tile Walls - Condensation", Technical Notes on Brick Construction, Brick Institute of America, McLean, Virginia, Notes 7C and 7D, 1968.

[3] Shipp, P. H., and Marchello, M. J., "What You Ought To Know About Air Barriers and Vapor Retarders", in Form & Function, USG Corporation, 1989, pp. 9-14.

[4] Kudder, R. J., Lies, K. M., and Hoigard, K. R., "Construction Details Affecting Wall Condensation", Symposium on Air Infiltration, Ventilation and Moisture Transfer, David Eakins, Chairman, Building Thermal Envelope Coordinating Council, Washington, D.C., 1986.

[5] ASHRAE Handbook of Fundamentals, American Society of Heating, Refrigerating and Air-Conditioning Engineers, Atlanta, 1989, Chapter 6.

[6] Complete Temperature Measurement Handbook and Encyclopedia, Omega Engineering, Inc., Stamford, Connecticut, 1985.

[7] Huang, P. H., "Humidity Sensing, Measurements, and Calibration Standards", in Sensors Magazine, Helmers Publishing, February 1990, p. 12.

[8] National Weather Service, "Surface Weather Observations", National Oceanic and Atmospheric Administration, U.S. Department of Commerce, Form MF1-10, Parts A and B, Daily.

[9] Hunderman, H. J., and Rose, W. B., "Humidity and Building Materials in the Museum Setting", in The Construction Specifier, Vol. 42, No. 10, October 1989, pp. 98-104.

Heinz R. Trechsel[1]

METHODS FOR IDENTIFYING SOURCES OF MOISTURE IN WALLS

REFERENCE: Trechsel, H. R., "**Methods for Identifying Sources of Moisture in Walls,**" <u>Water in Exterior Building Walls: Problems and Solutions, ASTM STP 1107</u>, Thomas A. Schwartz, Ed., American Society for Testing and Materials, Philadelphia, 1991.

ABSTRACT: The first issue in investigating moisture problems is the source of the moisture. Sometimes, visual inspection can determine whether liquid water or vapor is involved, active mechanisms, and season when problem occurs. Frequently, however, the issue is not so clear, and diagnostic tools must be used to establish the source and transportation mechanisms. This paper identifies some diagnostic tools and techniques which have been used successfully to determine moisture sources and transport mechanisms. It also discusses the locations, frequency, and duration of tests required. Examples of the application of the techniques are provided to demonstrate the application of the tools. Opportunities for innovative diagnostic technologies will be discussed, as well as the need for standard protocols for evaluating moisture problems for comparing results and to allow the establishment of a data bank as a basis for developing future guidelines for preventing moisture problems in buildings.

KEYWORDS: buildings, dew point, diagnostic tools, tracer gas, fan pressurization, infiltration, permeability, permeance, relative humidity, tracer gas, walls, water, moisture

---

Investigators are often given the task of determining the causes of moisture problems in a building, and then recommend remedial actions to repair previous damage and to prevent recurrence. The first step in such an investigation is the determination of the source or sources of the moisture. Without such determination, a rational and effective plan for remedial action cannot be suggested.

Sometimes the source is quite evident, but even when this appears to be the case, care should be exercised since appearances can be deceiving. What may seem to be a clear-cut case of rainwater leakage

---

[1] H.R. Trechsel Associates, Germantown, MD. Mr. Trechsel is also Criteria Manager at the Chesapeake Division of the Naval Facilities Engineering Command. Opinions expressed are those of the author and not necessarily those of the Naval Facilities Engineering Command.

may, after careful investigation, turn out to be in fact a condensation problem. Conversely, an apparent condensation or moist air leakage problem may actually have as its sources a combination of rainwater, construction moisture, and condensation of insufficiently dehumidified air. It is the purpose of this paper to stress the importance of identifying moisture sources through testing and measurement, that is by establishing a data base rather than by relying on visual observation or even "gut feeling."

Once the source of the moisture is determined, the preparation of remedial actions becomes a relatively simple matter, although such issues as the relative merits of providing better vapor retarders or increased ventilation will also need careful consideration. However, the availability of a data base will not only greatly simplify the decision making process, but will also raise the decision process onto the level of a straight-forward engineering discipline.

VAPOR OR LIQUID WATER

In most investigations, the primary question in determining the source of the moisture is whether the source is liquid water or water vapor, or a combination of the two. An example of a simple liquid water source may be a direct rainwater leak through a wall, where the water enters the wall at an opening, such as at a through-wall pipe, and manifests itself as a puddle on the floor below. An example of a simple condensation case is condensation on, and dripping from, a single glass window pane during cold weather. A combination of the two can exist when rainwater leaks through the exterior of the wall, evaporates, and condenses on a cool interior surface. Unless the entire mechanism of moisture source, transport, phase and phase change is known or correctly inferred, remedial actions can not be designed rationally. This can be accomplished only through the judicious use of appropriate diagnostic tools and effective test protocols.

DIAGNOSTIC TOOLS

Several appropriate tools for determining the sources of moisture are available. Some are used quite routinely; others have not been widely used.

There are no instruments on the market that will "identify" moisture sources. Rather, there are diagnostic tests available which, if used with understanding and care, can provide data that allow the determination of the mechanisms that may be active in a particular case, or which mechanisms can not possibly be active. Although there are many more tests, instruments and tools which may be useful at times, the five tests and data sources briefly discussed below have been employed successfully for the establishment of a useful data base:

- o   Air Infiltration Measuring Devices,

- o   Moisture Meters,

- o   Relative Humidity / Temperature Measuring Devices,

o   Water Vapor transmission tests,

o   Water Spray,

o   Weather Data.

In the experience of this investigator, the use of these devices, tests, and information sources, singly or in combination, will in most, if not all, cases allow the determination of the source of the moisture.

Air Infiltration and Movement Measurement

Two basic types of air infiltration measuring devices and methods exist currently: Those which use a tracer gas and the measurement or control of tracer gas concentrations at specified locations and time intervals, and those based on pressurization/depressurization.

Tracer Gas: There are several methods for determining air leakage into or air movement within building spaces, including wall cavities. The most simple one is the tracer dilution method in which a tracer gas is injected into the space and the decay of the tracer measured over time. ASTM E741, "Test Method for Determining Air Leakage Rate by Tracer Dilution," describes this method. Other tracer gas techniques for measuring air infiltration are based on constant injection rates and on constant concentration rates. The use of several tracer gases simultaneously allows the determination of air flow among several building spaces or cavities. At the present time, standardized test methods do not exist for any other than the dilution method. ASTM Subcommittee E06.41 is planning to cover all appropriate types of tracer gas tests. The primary drawbacks of tracer gas methods are the required electronic equipment and a competent operator. The advantage is that measurement can be made under actual air pressure conditions as they occur in the building. Since these conditions change with season and weather conditions, it is desirable to conduct multiple measurements at selected times and under different weather conditions.

Tracer gas tests are currently used primarily to characterize air flow into, out of, and within building spaces. However, these techniques can also be used effectively to determine air flow into, through, and within wall and other building component cavities. Although today no standard tests and protocols are available, with understanding of their use and limitations, tracer gas tests can be powerful tools for the investigator of moisture problems in buildings and walls.

Fan Pressurization: The second method is based on the application of an air flow generator, or fan, creating a positive or negative pressure inside a building space or enclosure, and measuring the air flow through the device. This is a simple and effective method for measuring air leakage into a building space under a given pressure difference. Results are dependent on the pressure applied, and on whether that pressure realistically simulates actual conditions in service. The results also do not take into account the location of the potential in-service air leakage sites. The method has not, to our knowledge, been used to measured air leakage into or air flow within wall cavities, except for determining air leakage paths. ASTM E779, "Test Method for Air Leakage by Fan Pressurization Devices" provides a standard for conducting tests by the pressurization method and ASTM E1186, "Air Leakage Site Detection in Building Envelopes" outlines methods for detecting the locations of air leaks. Because fan pressurization does not measure air leakage under actual service conditions, it is generally a less useful method than tracer gas for establishing the sources of moisture.

Moisture Content

The most reliable measurement method for moisture content is by weighing and oven drying sample materials. However, this method can not always be used, either because samples can not readily be obtained or because an immediate result is desired, thus no time is available for conducting a laboratory test off site. Tests by drying samples are discussed in ASTM C140, "Standard Methods of Sampling and Testing Concrete Masonry Units;" ASTM C472, "Standard Methods for Physical Testing of Gypsum Plasters and Gypsum Concrete;" ASTM D143, "Standard Methods for Testing Small Clear Specimens of Timber;" D1037, and "Standard Methods of Evaluating the Properties of Wood-base Fiber and Particle Panel Materials."

Moisture Meters are used for in-situ determination of the moisture content of building and other materials. Most moisture meters are based on the fact that the electrical resistance of materials change with a change in the moisture content. However, this change of electrical resistance is a function of the material and the meter can be used only with the material for which it was calibrated, or with appropriate tables. Moisture meters are most frequently used for measuring the moisture content of wood. They can, however, be used effectively with other relatively soft materials, such as gypsum board, if calibrated for them. Since the meters require the electrodes be inserted into the material to be measured, they are not readily used for hard materials such as masonry or poured concrete. It is possible to measure the moisture content of such hard materials in two ways. In one, the electrodes are inserted, or implanted into pre-drilled holes. In the other, a small probe (commonly called a "matchstick" probe), which in itself carries the electrodes is inserted into pre-drilled holes. In the latter case, the meter measures the moisture content of the probe only, but this content can be related to the moisture content of the material the probe is inserted into. Proper calibration of all devices for each application is mandatory for reliable results.

Air Temperature, Relative Humidity, and Dew Point

Air Temperatures: The measurement of air temperatures need not be discussed here in detail. Most meters used for determining relative humidity (see below) also provide temperature readings. As a rule, digital thermometers are more accurate and useful.

Relative Humidity: Sling psychrometers, measuring dry bulb and wet bulb temperatures simultaneously, were the traditional instruments used for measuring relative humidity. For the building investigator, the newer electronic devices which measure relative humidity and temperature simultaneously are more useful. Many types are on the market, both hand held and recording types. They typically have ranges of 0 to 100 percent RH and -4 to 150 °F (-20 to 60 °C), resolutions of 0.1 to 1.0 percent RH and 0.1 to 1.0 °F (0.1 to 1.0 °C), and accuracies from 2 to 4 percent RH and 0.5 to 1.0 °F (0.3 to 1.0 °C), all adequate for work in buildings. Most of these instruments have small (1/2 to 3/4 in (12 to 19 mm) probes well suited for collecting data through small holes in wall cavities. Recording devices are available which allow the recording of data over extended periods of time to establish trends and hourly and daily variations.

<u>Dew Point</u>: One reason for measuring temperature and relative humidity is to determine the dew point of the air in a room or in a wall cavity. The dew point can easily be determined from simultaneous readings of temperature and relative humidity on a psychrometric chart.

<u>Water Vapor Transmission Tests</u>

It is generally agreed (see ASHRAE Handbook of Fundamentals, Chapter 21 [1] for several references, ranging from Wilson and Garden in 1965 [2] to Handegord in 1979 [3]) that water vapor transport into and through the building envelope occurs primarily in the form of mass transfer by humid air. However, diffusion of vapor through materials does take place and can, in some cases, be the controlling factor. When this is the case, it is important to know the resistance of the various layers of materials and components to water vapor transmission.

Water vapor permeance tests would be most usefully done nondestructively in the field, but to our knowledge no such field test methods exist, and one is forced to collect samples and have the tests performed in the laboratory. ASTM E96, "Test Methods for Water Vapor Transmission of Materials" is most frequently used, but this test method has several drawbacks:

1. The method does not consider the effect of temperature.

2. The method is not easy to apply. The method assumes a perfect seal between the "cup" and the specimen. This is difficult to achieve. And the relative humidity, both ambient and in the cup, is difficult to maintain.

3. The test conditions of 0 % to 50 % or 100 % to 50 % RH rarely, if ever, correspond to the actual conditions to which a building material will actually be subjected over a prolonged period of time.

Recognition of the above problems (and others not mentioned), led several ASTM Committees to co-sponsor a symposium on "Water Vapor Transmission Through Building Materials and Systems" in 1987 in Bal Harbour, Florida. The proceedings of that symposium [4], contain many papers discussing these issues.

For example, Toas [5] reported on a round-robin test conducted by ASTM Committee C16 in 1985; it was found that the variation of results of tests on similar samples of mylar, conducted by twelve laboratories, varied significantly and varied even for tests conducted within individual laboratories. A broader discussion of other aspects of measuring water vapor transmission through materials was provided by Bomberg [6] during the same symposium.

Based on personal experience with costly tests and unreliable results, and because of the difficulties in conducing ASTM E96, it is strongly recommended to entrust the conduct of these tests only to laboratories with demonstrated expertise in such tests. It is also recommended that more than the three samples stipulated in E96 be tested to arrive at a more reliable average result.

Water Spray

Although not developed for determining water resistance of walls in general, ASTM E1105, "Field Determination of Water Penetration of Installed Exterior Windows, Curtain Walls, and Doors by Uniform or Cyclic Static Air Pressure Difference" has been used with success to determine water leakage under many different conditions. For masonry walls, the exposure time may need to be extended. Alternatively, ASTM E514, "Water Penetration and Leakage Through Masonry" can be adapted for field tests. This method also provides guidance for selecting exposure times.

Weather Data

It is obvious that weather plays a major role in most moisture problems, be it rainfall, relative humidity, temperature, and sometimes wind direction and speed. With the possible exception of rainfall and wind, maxima and minima data are not generally useful for moisture investigations. Averages and long term (seasonal) trends provide better results. It is not always easy to obtain, or infer, accurate weather information for a given location. However, local data can be obtained from even minor airports, and average weather data is readily available from the National Climatic Data Center in Asheville, NC. Their "Climatic Atlas of the United States" [7] contains a wealth of information in a convenient format. Design data for heating and cooling, and for energy calculation are also given in the ASHRAE Handbook of Fundamentals [8 and 9]. However, for purposes of moisture analysis, design data is less useful.

Since most moisture problems inconveniently occur in buildings somewhat removed from airports or other weather stations, the issue of interpolation of data for a particular location can be significant. Crow [10] provides examples for interpolation. As a general rule, data based on interpolation from stations at some distance should be used with caution and an awareness of their limitations.

NUMBER, FREQUENCY, AND DURATION OF TESTS

In applying the above tests, it must be understood that a single data point is better than none, but that a single point also can be misleading. Many moisture problems are seasonal, and it is not always clear which season is the critical one. For example, a major problem may occur in early summer and could be the result of the warm and humid summer condition or the delayed result of a cold weather condition. Thus, it is frequently desirable or even necessary to conduct tests over one season or over a full annual cycle. Building owners, usually the investigator's clients, generally want an early indication of where the problem is, and what can be done to "fix" it. Add to this that diagnostic tests can interfere with the operation of the building, the investigator is often under pressure to take shortcuts. It is wise to resist such pressures. They lead to faulty analysis, incorrect identification of causes of problems, inappropriate remedial actions, and a delay in the objective to truly solve the problem. Lack of a properly conducted data collection also can lead to lost law suits and result in the embarrassment of the investigator.

In preparing the plan for the investigation of a moisture problem, it is necessary not only to determine which tests to perform but also when and with what frequency the tests need to be performed. It is very important to develop a testing protocol which spells out the details of the tests and the times and frequency when the tests need to be performed. The preparation and availability of a detailed testing plan is also useful in persuading a client that the tests need to be undertaken.

LOCATION OF TESTS

As important as the number and frequency of tests are the locations selected for testing. It is normally not sufficient to take measurements only in one single location. The exact number of locations is dependent on the type of building, its design, and the nature and location of the problem. If the problem is clearly localized, it may be reasonable to assume that tests need be conducted only at the location of the problem. On the other hand, an all-pervasive problem may indicate that the same conditions occur throughout the building. Again it may be only necessary to conduct tests in one location. But in the former case, the problem may occur with only a short delay in additional locations, and in the latter case, there is no assurance that the same cause is responsible for the problems in all locations. Consideration should also be given to select locations in which no problems have been found. Data from such locations, when compared with data from locations which do have a problem, can identify differences in conditions which help identifying the causes for the problem. It is thus frequently necessary to collect data from a variety of locations, for example on low, intermediate, and high floors of a high rise building; on two or more orientations on walls with and on walls without balconies; on walls with and without windows; and in locations with and without problems. The investigator needs to determine those locations which are necessary to accomplish his objective in each case. As a general rule, it is prudent to collect more rather than less data.

If fewer locations are chosen, more care needs to be given to select "typical" locations. The temptation can be great to select convenient rather than truly typical locations. Convenient in this case may mean more accessible to the investigator or less disturbing to the client in the continuing operation of the building. Both the investigator and the client need to recognize the cost of such convenience in terms of a compromised, incomplete, or unsuccessful investigation.

EXAMPLES OF INVESTIGATIONS USING DIAGNOSTIC TOOLS

Below, we will discuss briefly three case studies. Two cases concern buildings located in warm climates. One of these cases deals with single-story, semi-detached masonry housing, the other is a mid-rise building. The houses are located one to two miles from the Gulf of Mexico, the mid-rise faced directly on an ocean beach. In both cases the investigations lead to similar conclusions. But while in one case the owner, the U.S. Navy, invested heavily in a multi-year study of a problem that affected potentially hundreds or even thousands of units and permitted extensive testing of many units, the owner of the midrise allowed only a minimal amount of field investigation. A thorough description of the former case was published by ASTM previously in STP 922 [11]. The study on the midrise building has not been published to date. The two studies are presented here only as examples illustrating the extensive use of diagnostic techniques on the one hand and their minimal use on the other. The third case concerns moderate cost homes in the northern Midwest. They too experienced serious moisture distress, and diagnostics were used to determine potential causes.

Masonry Housing

The buildings of this investigation experienced waterlogged gypsum wall board and severe mildew and mold growth on walls and on furnishings. An earlier study lead to the application of a coat of damp proofing on the interior face of the concrete block masonry wall and the replacement of the mineral fiber insulation in the furred space with a plastic foam board. While these remedial actions did seem to reduce the problem somewhat, within a few years the same conditions recurred. The previous remedial actions were based on the assumption that the problems were the result of insufficient dehumidification by the air-conditioning equipment in summer. Since the problems were apparent also in a house that had been unoccupied (with the air conditioning turned off) for an extended period, the investigators questioned whether air-conditioning during the summer month was the only or even primary cause of the problems and developed a strategy which could identify possible other moisture sources. Several diagnostic tools were employed.

On one single vacant unit, temperature and relative humidity were measured. Readings were taken every six to eight hours outdoors, indoors, and within concrete block cavities in several locations for two weeks. Four of these two-week sets of measurements were made approximately quarterly, for essentially a full year. In addition, an extensive occupant survey was conducted on some 85 units, and air infiltration and other tests were conducted at some 25 units.

It was found that during all but one set of readings (the one in November), the average dew point temperature inside the wall cavity was higher than the dew point temperature either inside the building or outdoors. From this it was inferred that the source of the moisture in the concrete block cavity could not be water vapor but had to be liquid water. This finding was also consistent with the results of water content tests of concrete masonry samples taken from the exterior of the block, from the block flange, and from the interior of the block. The web samples showed the highest moisture content. To verify that rainwater leakage was the, or at least one major source of the moisture in the wall, water spray tests were conducted. Indeed, these tests showed that copious amounts of water infiltrated into the concrete block cavities.

This investigation also demonstrates the significant contributions that tracer gas air leakage tests can make. The occupant survey and visual observations had shown that the preponderance of problems occurred in the bedrooms and not, as might have been assumed, in the bathrooms or in the kitchen areas. Tracer gas tests of individual rooms showed that with the windows closed (but bedroom doors open), the air change rate in the bedrooms was approximately 0.1 ACH and only one half the rate measured for the entire house. With the bedroom doors closed, the air change rate in the bedrooms was from near 0 to 0.1 ACH. The incidence of moisture problems thus correlated with low air change rates in the bedrooms. This suggests that the lack of ventilation in the bedrooms prevented the drying of the bedroom walls while the greater ventilation rate in the living room and kitchen areas may have promoted the drying of walls in these spaces.

<u>Mid-Rise Building</u>

This building showed distress similar to the houses above. In addition, mold growth on interior partitions was particularly puzzling Several theories were given relating to the causes of the moisture problems and the source of the moisture. One suggestion was that the major moisture source was infiltration of significant amounts of warm, humid air during the summer months. This air was thought to condense on the colder surfaces within and on the inside face of exterior walls, and on interior partitions. A more detailed analysis indicated, however, that the average dew point temperature of the outdoor air during the summer months was lower than the indoor temperature and that thus condensation of outside air could not occur within or on walls or furnishings at or near room temperatures. Furthermore, the incidence of mold did not correlate well with the measured indoor to outdoor pressure difference which was thought to be the cause for the movement of the warm, humid air into the hotel rooms.

Relative humidity and temperatures were measured with recording instruments indoors, outdoors, within the exterior wall cavity and within an interior partition abutting the exterior wall. Data were collected during several periods of a few days over a summer season only. As in the military housing example above, the dew point temperature within the exterior wall cavity was higher than the dew point of outdoor temperature, suggesting the presence of a rain leak. Again, water leakage tests confirmed that suggestion.

The readings within the interior partition indicated that the dew point temperature of the air in the cavity was significantly greater than that of the air in one of the adjoining rooms, only very slightly lower than that of the air in the other adjoining room, and somewhat lower than the dew point in the adjoining exterior wall. Since the detail at the intersection of the exterior wall and the interior partition indicated that the cavities of the two building elements were connected, this suggested that the source of the moisture in the partition was the same rain water leak responsible for the moisture in the adjoining exterior wall.

Moderate Cost Homes in the Northern Midwest

After several years in service, a number of these homes experienced moisture distress in the form of rotting plywood sheathing and mold growth on interior wall, trim, and window frame surfaces. Visual inspection indicated that the sheathing problems appeared most severe in the wall directly below the bathroom window and near the gable in the area of the bedrooms. Visual inspection also indicated that there was a building paper between the plywood sheathing and the siding, and that the thermal insulation blankets in the wall cavity was faced with a cover on both sides.

The mold growth on interior surfaces of exterior wall elements suggested a high indoor air moisture content. This seemed consistent with measurements conducted informally by several home owners.

Air infiltration measurements conducted by both blower doors and passive tracer (Perfluorocarbon) gas indicated that the houses were relatively tight (air leakage rates in the 0.2 to 0.6 ACH range). Water vapor permeance tests on samples of the building paper showed that the exterior building paper had a water vapor permeance in the range of 0.1 to 0.9 perms, and thus, according to ASTM C755, "Standard Practice for Selection of Vapor Retarders for Thermal Insulations," must be classified as a vapor retarder. Tests on the membrane facings of the thermal insulation blankets indicated that these facings had permeances greater than the exterior building paper. In some instances the facing of the insulation batt with the lower permeance was installed toward the exterior of the wall. The result was that the wall was less vapor permeable towards the exterior than toward the interior, trapping moisture inside the wall during cold weather.

This conclusion was consistent with a visual observation in the spring of 1987 on one house where on a four-foot section of the wall the exterior building paper had been omitted. The plywood sheathing in that place was completely dry, whereas it was dripping wet only a foot away where the building paper was in place.

Comparison of the Three Investigations

All three examples demonstrate the usefulness of diagnostic tools. The first case was carefully built on a significant data base established over a full year. The client in that case was fully supportive and willing to assume the cost and delay resulting from a major diagnostic effort. The second was driven by expediency. A more extensive data collection effort would have provided a more clear-cut case instead of one based at least in some degree on judgement. The third case falls somewhere in between: Financial constraints did not allow the careful diagnostic approach that would have been desirable, but careful analysis allowed the determination of the probable causes of the moisture problems. All three cases indicate the value of using diagnostic tools in wall failure analysis, specifically tracer gas and pressurization air infiltration and movement tests, temperature and relative humidity measurements, and water vapor permeance tests.

The second example also demonstrates how the lack of the use of available tools (such as tracer gas tests to determine air movement into and through the exterior wall and interior partition cavities), and data collected over only a short period and in only one or two locations can lead to, and require, a judgment call where it could and should have been a clear-cut engineering evaluation.

INNOVATIVE DIAGNOSTIC TOOLS

As the use of available tools increases, additional, innovative diagnostic tools will undoubtedly become available. ASTM D3017, "Standard Test Method for Moisture Content of Soil and Soil-Agregate in Place by Nuclear Methods (Shallow Depth)" is probably only a precursor of many more nuclear-based technologies that will be available for determining moisture content in the future. Similarly, the availability of remote controlled small sensors will reduce the need for frequent and disturbing visits by the investigator. This would lessen the resistance of building owners to extensive measurement programs. There are already available tracer gas systems that allow the measurement of air change rates averaged over extended periods of time. (See Dietz [12].) As mentioned above, multi-tracer techniques are being perfected and equipment may soon be available to permit all or most investigators to conduct such tests and evaluate results with confidence.

STANDARD PRACTICES AND PROTOCOLS

Just as important as the availability of diagnostic tools is the availability of standard practices and test protocols. They are needed as a guide for the newcomer to testing. They are even more essential to allow the comparison of tests performed by different persons, in different locations, and at different times.

ASTM Committee E06 has been in the forefront of standardizing test methods used in the building field. Furthermore, many of these methods concern building walls and wall elements such as windows, and several deal with air infiltration and air leakage detection. And E06 is not standing still! During its 1990 October meeting, a new task group was formed in Subcommittee E06.41 on Infiltration Performance to develop a practice for conducting air movement and gas diffusion tests in wall and other building cavities by tracer gas. The availability of such a practice could have been of use in the second of the examples outlined above. Subcommittee E06.55 on Exterior Building Wall Systems is in the proceeding to develop a protocol, or guideline, for investigating wall problems. Such a protocol will help investigators to evaluate moisture problems in walls, and could be the basis for establishing a data base on moisture problems in building walls.

Such a data bank, covering moisture related building failures would in the future permit investigators to compare individual test results with other similar constructions. In the opinion of this investigator, only the availability of significant and reliable data will eventually permit the establishment of realistic and achievable, yet high-grade performance criteria relating to new building walls, wall repairs, and wall retrofit. To establish the data base, however, it is necessary to conduct investigations in ways that generate measured data. This in turn requires the increased use and further development of field instrumentation and uniform protocols.

REFERENCES

[1] ASHRAE Handbook of Fundamentals, Chapter 21, American Society of Heating, Refrigerating, and Air Conditioning Engineers, Atlanta, 1989.

[2] Wilson, A.G. and G.K. Garden, "Moisture Accumulation in Walls Due to Air Leakage." Paper Presented at RILEM/CIB Symposium on Moisture Problems in Buildings, Helsinki, Finland, NRCC 9131, 1965.

[3] Handegord, "The Need for Improved Air Tightness in Buildings," Building Research Note 151, National Research Council of Canada, Ottawa, 1979.

[4] H.R. Trechsel and M. Bomberg, Eds., "Water Vapor Transmission Through Building Materials and Systems: Mechanisms and Measurements," STP 1039, American Society for Testing and Materials, Philadelphia, 1989.

[5] Toas, M., "Results of the 1985 Round Robin Test Series Using ASTM E96-80," Water Vapor Transmission Through Building Materials and Systems: Mechanisms and Measurements, STP 1039, H.R. Trechsel and M. Bomberg, Eds., American Society for Testing and Materials, Philadelphia, 1989,

[6] Bomberg, M. "Testing Water Vapor Transmission: Unresolved Issues," Water Vapor Transmission Through Building Materials and Systems: Mechanisms and Measurements, STP 1039, H.R. Trechsel and M. Bomberg, Eds., American Society for Testing and Materials, Philadelphia, 1989

[7] "Climatic Atlas of the United States," Environmental Science Services Admin., U.S. Department of Commerce, Asheville, 1983.

[8] ASHRAE Handbook of Fundamentals, Chapter 24, American Society of Heating, Refrigerating, and Air Conditioning Engineers, Atlanta, 1989.

[9] ASHRAE Handbook of Fundamentals, Chapter 28, American Society of Heating, Refrigerating, and Air Conditioning Engineers, Atlanta, 1989.

[10] Crow, L.W. "Study of Weather Design Conditions," ASHRAE RP 23, American Society of Heating, Refrigerating, and Air Conditioning Engineers, Atlanta, 1963.

[11] Trechsel, H.R., Achenbach, P.R., and Conklin, S., "Field Study on Moisture Problems in Exterior Walls of Masonry Housing on the Coast of the Gulf of Mexico," Thermal Insulation: Materials and Systems, STP 922, F.J. Powell and S.L. Mathews, Eds., American Society for Testing and Materials, Philadelphia, 1987.

[12] Dietz, R.N. et al., "Detailed Description and Performance of a Passive Perluorocarbon Tracer System for Building Ventilation and Air Exchange Measurements," Measured Air Leakage of Buildings, STP 908, H.R. Trechsel and P.L. Lagus, Eds., American Society for Testing and Materials, Philadelphia, 1986, pp. 203-264.

G. Gabriel Cole and Thomas A. Schwartz[*]

ESTABLISHING APPROPRIATE FIELD TEST PRESSURES FOR INVESTIGATION OF LEAKAGE THROUGH THE BUILDING ENVELOPE

REFERENCE: Cole, G. G., and Schwartz, T. A., "Establishing Appropriate Field Performance Test Levels for Investigation of Leakage Through the Building Envelope", Water in Exterior Building Walls: Problems and Solutions, ASTM STP 1107, Thomas A. Schwartz, Ed., American Society for Testing and Materials, Philadelphia, 1991.

ABSTRACT: Most performance standards for building envelope components, such as those for certified windows, have been developed for evaluating new products or for helping assure the quality of mass produced products. There are, however, few standard procedures for evaluating performance problems with installed wall systems. Many investigators have attempted to apply performance standards for new assemblies to problematic buildings; and this often leads to the wrong conclusion regarding the cause of building leakage.

This research explores the appropriate use of water test procedures for field investigation of leakage through building walls. We explore both the type of testing and appropriate criteria for judging the contribution to leakage from various components of the exterior envelope. We also discuss a methodology for developing test parameters based on actual weather records and surrounding topography. These exposure conditions can be used to formulate test pressures that can be applied to elements of the building envelope during water testing to more closely simulate the past exposure conditions that these wall components experienced during leakage events.

KEYWORDS: water testing, wind effects, pressure differential testing, leakage, window performance standards, weather exposure

INTRODUCTION

One of the most challenging elements of building diagnostic work is the identification of sources and paths of water leakage. This evaluation is complicated by the constant variability of weather conditions that give rise to leakage, as well as by the increasingly complex combinations of materials in modern building facades. Evaluators of building leakage must be thorough and persistent in their approach to ensure that all potential sources of leakage are evaluated.

---

[*] Mr. Cole is a senior engineer and Mr. Schwartz is a principal at Simpson Gumpertz & Heger Inc., Consulting Engineers, 297 Broadway, Arlington, MA 02174

It is also important that water test work conform to standard methods so that tests on different elements or in different areas of the building are comparable. Such tests can then be used to rank the contribution of these various building components and subsequently prioritize the remedial work. In recent years both ASTM and the American Architectural Manufacturers Association (AAMA) have developed field test methods for water testing installed windows, doors, and curtain walls. None of these field test procedures provide a method for establishing rational test pressures to approximate the weather conditions responsible for actual leakage of wall assemblies in service.

All window units certified by AAMA and the National Wood Window and Door Association (NWWDA) must be capable of passing a water infiltration test at a specified air pressure differential to obtain approval. In recent years, those involved in the evaluation and remediation of building envelope failures have frequently applied these standards to in-situ construction. As a result, the most prevalent misuse of standard water tests that we have encountered involves testing of window assemblies on existing structures with an air pressure differential across the unit specified for window "certification." In some cases, failure of a window to pass these tests has led an investigator to conclude, incorrectly, that the window unit is the sole cause of building leakage. The use of "certification" test pressures for evaluating reported leakage problems in the field is often inappropriate for the following reasons:

- The "certification" test pressures are generally higher than the pressures that existed on the window or door elements when actual leakage occurred in the field.

- These tests require application of a "certification" test pressure to the test specimen for either a constant 15 minutes or four cycles of 5 minutes depending on the test selected; such durations generally do not model the fluctuating effects of wind.

The pressure magnitude and load duration requirements prescribed by the "certification" test standards generally represent severe "proof test" parameters used for prototype development, quality control during manufacture, and field verification during the construction process.

The challenge to building investigators is the development of a rational approach to assigning water test pressures for field use to replicate the conditions responsible for actual leakage.

FIELD TEST METHODS FOR ASSESSING LEAKAGE

AAMA has recognized the need for field testing of newly installed windows and curtain walls for several years, and, has progressively developed the following standards for field testing:

- 501.2-83: Field Check of Metal Curtain Walls for Water Leakage [1]

- 501.3-83: Field Check of Water Penetration Through Installed Exterior Windows, Curtain Walls and Doors by Uniform Air Pressure Difference [1]

- 502-90: Voluntary Specification for Field Testing of Windows and Sliding Glass Doors [2]

ASTM has also responded to the need to develop standards for field testing by developing standard E 1105-86: Standard Test Method for Field Determination of Water Penetration of Installed Exterior Windows, Curtain Walls, and Doors By Uniform or Cyclic Static Air Pressure Difference.

All of these tests are, however, geared toward assessment of specific elements and are intended primarily for quality assurance verification, during the construction project. The basic water test procedures described in these standards can be successfully employed in the evaluation of leakage problems on existing buildings. Since some of the test procedures are product specific and geared toward testing only window units, investigators must be careful to modify these tests appropriately and consistently when assessing the performance of the exterior wall assembly as a whole.

GENERAL GUIDELINES FOR BUILDING WALL LEAKAGE INVESTIGATIONS

Our past testing experience indicates that there are three broad categories of building leakage. Each of these leaks require a different test methodology to replicate the leakage in the field, as described below:

- Open defects, particularly if they are at a key sealant joint or flashing will generally leak with water spray only after a limited period of time (e.g. 15 minutes).

- Smaller defects, metal-to-metal joints in fenestration, and window or door sill assemblies often require application of a pressure differential to produce leakage.

- Massive wall systems, such as brick veneer and precast concrete often require prolonged application of water (e.g. one or more hours) to produce any leakage.

In this paper we will explore an appropriate procedure for assessing leakage that results from a pressure differential, such as a wind-driven rain. Our previous experience indicates that the wall should first be evaluated for open defects prior to performing pressure differential testing. Therefore, a practical approach to identifying water leakage includes, as a minimum, the following steps:

- Obtain leakage records for the building, and interview tenants and maintenance personnel to identify typical leakage locations. Obtain precise dates when leakage occurred, if possible. If you cannot obtain precise leakage dates, attempt to determine the leakage frequency (e.g. once or twice a year vs. every time it rains).

- Carefully observe interior finish surfaces and document evidence of stains and damage from actual leakage. The water penetration observed during subsequent testing should occur at these areas. If the leakage occurs at areas with no previous staining, this may indicate improper test parameters.

- Conduct similar water tests of all components that might contribute to the leakage. It is generally appropriate to first use a water spray rack in accordance with ASTM E II05-86, without using an applied pressure differential to simulate the effect of wind. The zero pressure differential spray rack test helps to establish the principal contributors to leakage, and the relative contribution of the various components. This comparative analysis aspect of this investigative method is invaluable. It helps avoid the problem of focusing on elements that may play a minor or insignificant role while overlooking the major contributors.

- If the precise point of water entry and its path to the interior is not confirmed by the spray rack test, a hand-held nozzle, as specified in AAMA 501.2-83 may be used to pinpoint the water entry location and water leakage path. The specification requires a water pressure of 30 psi (206 kPa), but our experience indicates that pressures of 0 (pour test) to 10 psi (69 kPa) are sufficient test levels to identify leakage paths without creating new leaks which would probably not occur under actual weather conditions.

- If zero pressure differential water tests do not show the sources of leaks that cause the interior stains or damage, and if the building has a severe weather exposure, it may be necessary to conduct water tests combined with a pressure differential across the wall to simulate wind-driven rain.

The pressure differential can be created either by applying a positive pressure to the exterior of the building, or by applying a negative pressure to the interior of the building. By far the most common test utilizes the interior chamber, which in its simplest form can consist of a sheet of plastic film taped to the interior of a window frame.

If pressure differential testing is used, the test chamber must be carefully designed, so that not only the windows, but surrounding elements are tested in a like manner. Our experience indicates that windows, window perimeter joints, and all types of surface-sealed wall systems are susceptible to increased leakage under a pressure differential.

Leakage through cavity walls is not generally influenced by an interior-applied air pressure chamber, unless provisions are made to depressurize the cavity area as well. This is difficult if not impossible on most cavities because a discrete chamber cannot be formed. Our experience indicates that leakage associated with these walls often requires prolonged application of water spray, without a pressure differential. Determination of the parameters for this type of test is important, but is beyond the scope of this paper.

ESTABLISHING APPROPRIATE AIR PRESSURE DIFFERENTIALS FOR WATER TESTING TO DETERMINE THE CAUSE OF PAST LEAKAGE

The critical element in designing a pressure differential test and in interpreting the significance of the results is the selection of an appropriate differential air pressure to be applied to the specimen. The pressure differential should approximate the actual wind pressures that occurred during leakage. To develop the pressure differential, we have used a procedure based on an analysis of available wind speed data, from which we can approximate the exposure pressure for periods of previous leakage.

Selection of Wind Speed Data

The first task in determining a water test pressure is to research the weather conditions under which actual leakage has occurred. Ideally, we would like to have a comprehensive record of weather taken at the exact location of the building, since weather in general and wind in particular are highly variable phenomena. Unfortunately, most buildings do not have their own weather station and we must, therefore, obtain the weather records from a nearby reporting station. These stations are generally located at major airports. We generally select the weather from the station which is closest in proximity, with the least amount of topographical change from the building. Since all airports are flat with few obstructions, some topographic change to the building site is inevitable. Data from a station separated by many abrupt topographic changes may, however, be substantially different from that at the site.

The National Oceanic and Atmospheric Administration (NOAA) publishes a variety of data collected at weather stations throughout the United States. NOAA publishes selected portions of the data collected at each of these stations in a monthly pamphlet titled "Local Climatological Data" [3]. This publication provides a daily summary of average and extreme conditions, including rainfall and extreme winds. Unfortunately, this document does not address whether or not the maximum wind occurred during a period of rain. Most weather stations actually record the weather conditions on an hourly basis, and more frequently during periods of significant weather activity to determine the peak values. This data is not reproduced in "Local Climatological Data." A complete listing of all data recorded at individual stations, known as "Surface Weather Observations" is available in microfiche form from the National Climatic Data Center in Asheville, North Carolina. This source allows a reliable correlation between extreme winds and rainfall.

In most cases the weather occurrences at the weather station and at the building will be nearly identical when averaged over a lengthy period of time, such as a year. Therefore, it is important to develop a "pool" of values and establish a distribution of wind-driven rain prior to selecting a test pressure. Once this pool is established, compare the maximum wind-driven rains with the building leakage records to determine if a correlation exists between reported leakage and the most severe wind-driven rains. If such a correlation does not exist, the building should be critically examined to determine if the method is in error due to any of the following factors:

- Wind direction: Some structures are only subject to leakage during wind-driven rains, when the rains are from a specific compass direction. If this is the case, the "pool" of wind-driven rain values should be modified to reflect wind direction. The maximum values for the "leakage direction" should then be checked against the reported leakage to verify a correlation.

- Weather station: In some cases, unusual topographic conditions may contribute to very localized weather phenomenon which can cause substantial differences between the weather observed at a weather station and that at the site. The probability of this occurring increases with distance from the weather station. This can be most easily addressed by reviewing records for other weather stations in the vicinity. If no other stations are nearby, the only other alternative is to mount a wind gauge on the building for several weeks and attempt to correlate the average values with those at the airport, or use these values to compute the test pressure if any leakage events occur during this time period.

- Type of leak: The leak may not be related to a pressure differential, but may be related to the duration of rainfall. In this case, pressure differential testing is not appropriate.

If building leakage records are not sufficiently detailed to indicate specific leakage dates, we generally develop a pressure differential based on the 3 to 5 most severe storms during the service life of the wall. This method is also subject to the errors noted above, and the test results should be closely monitored to ensure that the damage-causing leaks are replicated by this procedure.

Weather stations generally report two values for maximum wind speed--gust and one-minute average. Gusts are typically very short in duration and, as such, are not likely to significantly influence the leakage response of most wall elements, since the duration is so short that it simply will not move the water through the wall system to the point that it is considered leakage. The one-minute average corresponds to the average speed of wind passing a single point for a one minute period. This period is an excellent indicator of maximum weather activity, and is the basis for the structural loading criteria described in "Minimum Design Loads for Buildings and Other Structures" (ANSI A58.1), which governs most wind-related structural design in the United States. The wind spectrum, which represents the distribution of wind occurrence with respect to duration, reaches its maximum for frequencies between 20 and 100 seconds [4]. Since most wind events which occur are, on average one-minute long, it stands to reason that the maximum one-minute average will contain the most severe exposure conditions for a wall system. Use of this value does, however, require some modification of the standard testing procedure as described in ASTM Test E1105-86, to include a cyclic application of 1 minute duration, rather than the 5 minute duration recommended in the standard. In our work, we generally alternate one-minute cycles of pressure application with one-minute rest periods for 20 minutes. This is a more accurate approximation of the characteristics of a severe, wind-driven rain.

Method for Determining Test Pressures

Once the wind data for periods of leakage from the nearest weather station has been collected and analyzed, it is necessary to translate the wind data into a test pressure. The test pressure can be calculated based on the terrain conditions and height of the building. Prior to commencing this work, it is important to examine the building critically to determine if there are any peculiarities of surrounding terrain which may cause unique wind loadings, or if there are any peculiarities in operation of the building (e.g. the mechanical system may be designed to maintain a constant negative pressure on the interior of the building).

It is impossible to determine the precise wind pressures occurring at the site without instrumenting the building or performing a comprehensive wind tunnel study. Even these techniques are subject to error as a result of the highly variable phenomenon of wind action and the difficulty of assessing the actual pressures on the building facade.

On several projects, we have calculated what we consider to be a conservative value for the pressure imposed by wind using the method described in AAMA publication CW-11: "Design Loads for Building and Boundary Layer Wind Tunnel Testing" [5]. This publication contains a section on establishing design windloads which was developed for structural design purposes, but which we have adapted for use in developing a threshold pressure for water leakage given the wind data collected coincident with rain.

There are limitations to this method which should be fully understood before employing it in the development of a test program. AAMA provides basic wind speeds to use in calculating wind pressures, but they also permit substitution of other wind velocity values if the following applies:

- accepted extreme-value statistical analysis procedures have been employed in reducing the data,

- due regard is given to the length of record, averaging time, anemometer height, data quality, and terrain exposure, and

- the basic wind speed is not less than 70 mph (113 km/h).

Our procedure complies with all of these provisions, with the exception of the latter. Our previous experience indicates that, on many buildings, the three or four most severe periods of wind accompanied by rain can be as low as 25 to 35 mph (40 to 56 km/h), which is significantly less than the minimum requirement stated above. The requirement for using a minimum wind speed of 70 mph (113 km/h) exists both to ensure that minimum structural design criteria is met, and because the technique becomes less accurate at low wind speeds. The structural point is of no concern here, since we are not developing these values for use in structural design. The accuracy of the method is of great concern, since we are attempting to approximate the actual wind pressure on the structure, during periods of leakage. The method is not accurate for very low wind speeds (i.e. on the order of 5 mph; 8 km/h) because these speeds are generally the result of ground level phenomenon, which may vary substantially between the site and the weather station. High winds (i.e. on the order of 70 mph; 113 km/h) are typically the result of air movements in the upper atmosphere which do not vary substantially over distances of several miles. The intermediate winds which most often result in water leakage fall between these two extremes. Therefore, this technique will be in error on some occurrences. Since we are developing a pressure value based on the average of several wind speeds, these errors will tend to balance out and should not significantly change the pressure value.

The procedure developed by AAMA requires determination of the following items prior to calculating the "design pressure":

- Exposure Conditions (α, z_g): AAMA designates four separate exposure conditions--Large city centers (exposure A), urban and suburban areas (exposure B), open terrain (exposure C), and coastal areas (exposure D). Most airports are located in areas designated exposure C.

- Interior Pressure Coefficient (GC_{pi}): In addition to the positive pressure, which is imposed on the windward side of the building, negative pressure is created inside of the building due to suction on the leeward side of the building. For most structures, the negative pressure on the interior of the building is equal to 25% of the positive pressure on the exterior. If the building has many more windows or other openings on one face than it does on the others it may be necessary to modify this value.

- External Pressure Coefficient (GC_p): Generally this will be 1, unless the desired test area is near the edge of a wall, in which case the positive pressure on the exterior face will be increased. This particular value should, however, be used cautiously since it includes gust response factors, which are intended to approximate peak wind effects of only a few seconds duration. Therefore, use of these values is only appropriate in situations where leakage starts relatively shortly after application of the test pressure, once the system has reached equilibrium (e.g. it may take 5 or more cycles for leakage to occur, but once that leakage starts it reappears with each cycle shortly after the pressure application).

- Importance Factor (I): This is intended to increase the factor of safety for critical structures, but consideration of this is not necessary to develop the pressure for water testing.

Once these basic values have been obtained, the test pressure can be calculated by using the following steps:

1. Determine K_z for the airport and for the building site according to the following equation, where α and z_g are constants based on the exposure conditions, and z is the height of the proposed test area:

$$K_z = 2.58 \left(\frac{z}{z_g}\right)^{\frac{2}{\alpha}} \text{ for } z > 15 \text{ ft}$$

2. Calculate velocity pressure according to the following equation, where V is the one-minute average wind speed:

$$q_z = .00256 \ K_z \ V^2 \ \text{lbf/ft}^2$$

3. Finally, calculate the total exposure pressure according to the following equation, where q_z is evaluated at height z from ground level and q_* varies according to the face of the building under evaluation.

$$p = q_*(GC_p) - q_z(GC_{pi})$$

Uncertainties Associated with this Method

This method has proved to be an efficient procedure for calculating exposure conditions, since weather records are readily available for all regions of the United States. Our past experience shows good correlation between the leakage produced through or around wall components when using the test pressures developed by this method and the physical evidence of past leakage at the building (ie. interior stains and damage). There are, however, many uncertainties associated with this method. This method should not be considered an absolute method of determining pressure at a specific point in time, since the airport may be some distance from the site, but rather should be considered an approximate method to determine average exposure conditions.

Much of the uncertainty inherent in this method stems from the fact that there is very little research present concerning wind-driven rain or low velocity winds. Most wind research is related to extreme winds and is geared solely toward developing structural design criteria.

There is a need for considerable original research on the subject of wind-driven rain, and this paper does not attempt to answer all of the many questions that exist. Research is needed to determine if:

- Airport wind values approximate values at buildings in surrounding areas, given appropriate modifications for changes in terrain. Since the airport is physically in a different place there will not be a continuous matching of values, but the key is whether or not the average values, upon which our calculations are based, match.

- Calculated pressure values approximate actual pressure values. This research should also include development of better guidelines to determine the types of buildings for which this procedure is most accurate. This will require careful instrumentation of several buildings and monitoring over an extended period of time.

- Development of a wind-driven rain index for various areas of the country, based on peak winds coincident with rain.

SUMMARY

We have proposed a method for determining appropriate air pressure differentials during water testing for field investigation of leaky buildings, and we have briefly mapped out the areas in need of future research to define the uncertainties associated with this method. This method represents a rational way to approximate exposure conditions.

In using this method building investigators should keep in mind its primary role, which is to replicate leakage which has been noted by building occupants. The success of any testing method is contingent upon the accuracy with which it can reproduce this leakage.

REFERENCES

[1] Architectural Aluminum Manufacturers Association, "Methods of Test for Metal Curtain Walls," AAMA publication 501-83, AAMA, Chicago, IL, 1983.

[2] Architectural Aluminum Manufacturers Association, "Voluntary Specification for Field Testing of Windows and Sliding Glass Doors," AAMA publication 502-90, AAMA, Des Plains, IL, 1990.

[3] National Oceanic and Atmospheric Administration, "Environmental Information Summaries," National Climatic Data Center, Asheville, NC, 1988.

[4] Simiu, Emil, "Wind Spectra and Dynamic Alongwind Response", <u>Journal of the Structural Division</u>, Vol. 100, No. 6, 1974, pp. 1897-1910.

[5] Architectural Aluminum Manufacturers Association, "Design Windloads for Buildings and Boundary Layer Wind Tunnel Testing", AAMA Aluminum Curtain Wall Series No. 11, AAMA, Des Plains, IL, 1985.

Jeffrey C. May and Jan M. Vassiliades

TRACING ROOF AND WALL LEAKS USING ALTERNATING ELECTRIC FIELDS AND VAPOR DETECTION

REFERENCE: May J. C., and Vassiliades, J. M., "Tracing Roof and Wall Leaks Using Alternating Electric Fields and Vapor Detection," Water in Exterior Building Walls: Problems and Solutions, ASTM STP 1107, Thomas A. Schwartz, Ed., American Society for Testing and Materials, Philadelphia, 1991.

ABSTRACT: Exterior water can penetrate a building and move in either continuous or discontinuous flow. In the case of continuous flow, from point of entry to point of observation, surfaces of construction materials are wetted and may provide a path of adequate conductivity to produce a detectable alternating electric field when a voltage is applied. Where this is the case, a suitable a-c source and detector can be used to determine the source of water entry.

For cases where water flow is discontinuous, detection of vapor from water mixed with organic solvent can assist in tracing water movement and source of origin. Commercial equipment is available to measure, non-destructively, moisture behind and within non-conducting substrates such as wood and exterior foam panels. Using these methods one can often unambiguously locate the source of a water leak.

KEYWORDS: moisture detection, leak detection, alternating electric field, organic vapor

---

Porous materials such as wood have a lower electrical resistance when moist than when dry; this is the basis of operation for the standard direct current (d-c) pin-probe meter. Similarly, if an alternating current (ac) can flow through a material that is partially conductive due to its moisture content, the a-c field can be detected, and remote detection of the moisture is possible.

One can also detect increased a-c conductivity due to moisture content by combining the emitter and detector into a single instrument; such instruments are available and their use will be described.

Jeffrey C. May, M. A. is a building consultant for J. May Home Inspections/JMHI, Inc., 94 Wendell Street, Cambridge, MA 02138; Jan M. Vassiliades is an architect at M.B. Associates, Technical Support, 80 Lincoln Street, Boston MA 02111.

In construction materials, detecting the presence of moisture is generally easier than locating the source of the moisture. Typically, a suspected source can be leak-tested with water and observations made to determine if moisture appears at the anticipated location. In such leak testing, unfortunately, the movement and volume of water may not be detectable; however, with the addition of volatile marker chemicals, detection of the marker may provide positive proof for water entry. Relatively inexpensive instruments are available for the detection of organic vapors and their use will be discussed.

REMOTE ELECTRODE AND DETECTOR

Often, the source of water entry in a wood-clad exterior is an improper flashing detail. A non-grounded metallic flashing in wood siding can be employed as an "electrode" by attaching a low voltage source of ac to it; if there are concealed, moist construction materials around the flashing, they may provide a continuous conducting path and the extent of the moisture will be delineated by the extent of the electric field.

In actual practice, one must limit the possible current that can flow through the testing apparatus; if there are grounded conductors in poor electrical contact with the flashing, arcing and possibly ignition of materials in contact with the electrical gap could result.

One can limit the current flow by adding a resistance in series so that the maximum current flow is some fraction of an ampere. (Addition of a large resistance in series also eliminates the possibility of an accidental electric shock.) In practice, the wire from a variable voltage source is attached to one lead of one or more resistors so that the total resistance is over 60,000 ohms; the other resistor lead is attached to the wire that serves as the voltage source for the electrode. (A 5-watt tungsten filament light bulb can also be used as the resistance, but more current will flow if grounding is present.) A suitable alligator clip is attached to the end of the wire (See Fig. 1).

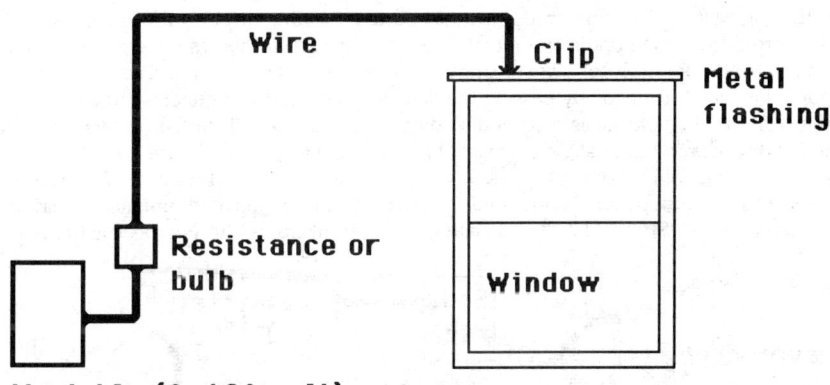

Fig. 1--Setup for using ac field tracing. TIF a-c field detector not shown.

One device for detecting an a-c field is the TIF 100 AC Voltage Detector (TIF Instruments, Inc., Miami, Florida), an inexpensive hand-held unit that operates on a 9-volt battery and emits a pulsed tone; in the presence of an increasing a-c field, the pace of the pulse increases as the field strength increases until the device emits a continuous tone. The unit "antenna" is held close to the exterior wall and the surface around the flashing is "scanned" with the detector. If an electrode is grounded the field cannot be detected around the electrode.

During a leak test, water is added around a suspect flashing; the spread of the water may be detected by the change in size of the a-c field. Note that the voltage source should not be left on continuously since current flow may alter the conductivity in the vicinity of the electrode. (The setup can be used to single out the leaves of a particular vine in a large mass of vines by attaching the electrode clip to the base of one stalk; all the leaves on that stalk alone emit the a-c field.)

Example A

A home owner was troubled by paint failure on exterior clapboards and trim around a sliding door. Several feet away from the door was a wood gutter that terminated perpendicular to the wall at the sheathing. There was a possibly leaky aluminum "end-cap" flashing. An a-c electrode was connected to the flashing and the area of field transmission tested with the TIF a-c field detector by pressing the antenna against the siding; the extent of transmission was large and denoted with chalk on the clapboards. Water was added to the gutter; shortly afterwards, the a-c field was delineated again with chalk. The field was detected at a greater distance from the source, suggesting moisture flow behind the siding.

The above setup was being developed when moisture detecting equipment manufactured by Tramex, Inc. (Bedford, MA) came to our attention. Tramex manufactures the "Side-Kick," a portable electronic device that uses a 9-volt battery to generate and detect an a-c voltage. The unit is equipped with two detachable leads: one lead has a clamp at the end for attachment to the "ground" of a structure; the second lead, connected to the detector, has a probe consisting of a long handle with a flat steel conductor. The principle use of the "Side-Kick" is to detect moisture under EPDM roofing.

In practice, the "ground" clamp emitter is attached to a flashing or other roof penetration that passes through the membrane and metal decking (Fig. 2). In this way, the a-c voltage is applied to the entire roof deck surface - if conductive- below the membrane. The surface of the membrane is then "scanned" by passing the detector over (and in contact with) the membrane surface. The detector is designed so that it can maintain electrical contact with the membrane as it is moved under stone ballast. At areas where there is moisture under the membrane, the a-c signal is stronger because it is conducted through wet materials below, such as insulation and/or composition board, and the partially conducting, carbon-black containing EPDM. (Note that the "Side Kick" detects actual a-c current flows and not electric fields.)

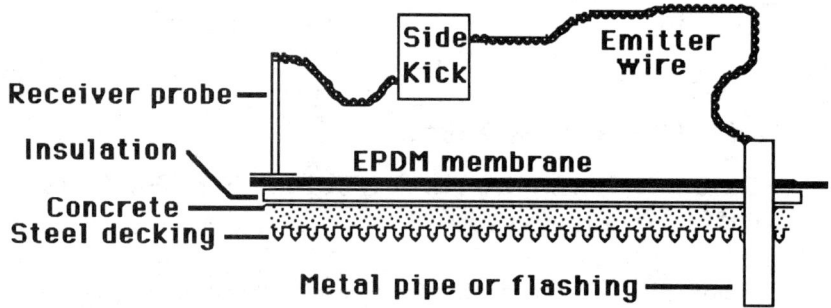

Fig. 2--Schematic for Tramex "Side Kick." Probe must be in contact with membrane to detect current.

ELECTRODE AND DETECTOR IN FIXED PROXIMITY

Tramex, Inc. also manufactures several electronic meters employing a-c signals for remotely measuring moisture. One moderately-priced hand-held meter -the "moisture encounter"- operates on a 9-volt battery and has two flat conducting foam rubber pads at the back of the case in close proximity. The emitting pad sends out an a-c signal that is received by the other pad. In practice, the unit is placed flat on a surface to be tested; the "conductivity" (current flow between the pads) of the sample is indicated on a meter. There are two relative moisture scales (for roofing and for masonry) and one "percent scale" that reads actual moisture content in wood.

The field emitted by the meter pad penetrates wood (and other materials) to a depth of about 1.25 cm and therefore can detect moisture below or behind the surface of a non-conducting material. For comparison, a pin probe meter might find normal moisture in the partially dried, exposed surface of a clapboard, whereas the "encounter" will detect wetness at the back of the board resulting from a leaking flashing. The non-invasive feature of the pads allows one to scan an entire wall without damaging the surface in any way.

The "encounter" will also detect moisture behind 65 mm ceramic tiles; another Tramex moisture detector- the "Leak Seeker"- intended for built-up roofing, has over four times greater depth of penetration and can be used to detect moisture behind or in foam panels that are part of exterior insulated foam system (EIFS) walls.

Example B

A medical practice was forced to abandon one of its examination rooms when patients complained of asthma and allergy problems in the room. The carpet had been discarded and the floors cleaned repeatedly with bleach, with no improvement of the moldy odor. The wall surfaces were scanned for moisture by sliding a Tramex "moisture encounter" along the walls. At a location under one window, elevated moisture was noted. At the exterior, an improperly fitted downspout was allowing water to drain onto the exterior window sill; the sill was cracked and allowing water flow into the wall. Subsequently, the interior drywall was removed; the fiberglass insulation was heavily contaminated with mold.

LEAK TESTING WITH ORGANIC VAPOR DETECTION

The TIF 8800 (TIF Instruments, Inc.) is a moderately priced, hand-held rechargeable combustible gas and organic vapor detector with adjustable sensitivity down to about 50 parts per million. The TIF 8800 has a flexible tip containing a semiconductor detector, the resistance of which varies with the equilibrium concentration of ambient gases. The unit produces an audible pulsed tone; if there is an increase in the concentration of any of the gases or vapors to which the unit is sensitive, the pace of the pulse increases until the device emits a continuous tone.

When a likely leak-candidate is chosen, water containing no more than 10% alcohol can be added to the site. If there is vapor detection at the remote site (and other more obvious means of vapor travel can be eliminated), then the source may have been located. In this way, leaks in EPDM membranes and masonry have been corroborated; and conversely, suspected leakage sites eliminated.

Obviously, leak testing with alcohol should not be undertaken in the vicinity of likely ignition sources such as electrical or combustion equipment, nor should large quantities of alcohol be employed.

Example C

At a university science center, water had been leaking from the entry hall ceiling since construction 18 years ago. Much to everyone's embarrassment, the leak continued after the new million dollar roof was installed. The entry ceiling consisted of hollow core, precast concrete planks with insulation and membrane roofing above; after heavy rain, water leaked for days from a hole in the bottom of one plank about 8 feet from the corner of a large, glazed atrium.

Water doped with alcohol was added at the roof into a torn (flashing) riglet at a location that fortuitously coincided with the end of the dripping precast plank. No alcohol vapor was detected at any time at the interior near the ceiling. Water doped with alcohol was then added from the exterior to the glazing system; vapors were detected at the hole in the bottom of the precast plank. Water had been leaking into the glazing system and running under the insulation along the top of the precast planks. Repairs to the glazing seals and reopening of the caulked weep holes terminated the leakage.

Example D

A painter was called back to a job when recently-painted clapboards started to peel between two bathroom windows. Shower moisture was suspected as the cause, but the stall shower walls were intact tiles set in mortar. There was a large center chimney at the roof ridge. In the vicinity of the roof edge, the Tramex "encounter" measured lower relative moisture in the fiberglass shingle roofing to the right and left of the chimney. Water doped with alcohol was added to a crack in the concrete chimney cap. Vapor was detected in moisture that appeared at the roof edge starter course. Apparently, no chimney flashing was installed when the new roof was placed over the old one; water was running down the chimney masonry into the space between the roof layers, down the roofing singles and into a crack of a starter course wood shingle; from there, water found its way into the wall sheathing.

Example E

A top floor condo owner was upset by ceiling efflorescence, stains, and peeling paint under a new EPDM roof. At a location above the site of staining, there was a parapet wall with membrane flashing containing a small hole (about 2 mm). Potentially large amounts of drainage rain water flowed over the hole; water doped with alcohol was injected with a syringe into the membrane hole. A 3 cm hole at the damaged interior plaster ceiling was made into the attic crawl space and the TIF 8800 probe inserted; organic vapor was detected by the unit.

CONCLUSION

No remote test method can distinguish between exterior water entry and condensation; and test results of moisture or leak testing methods are often not definitive. For example, the conductivity of a sample measured by a Tramex "encounter" (or even a pin-probe meter) may be the result of contaminant salts or metal and not simply sample moisture. The TIF vapor detector may pick up and respond to rooftop combustion fumes from a nearby chimney, or even fragrance from a bathroom or dryer vent.

It is common to find that numerous attempts have been made to "repair" a leak. Generally, one of the attempts will prove successful, though it is not uncommon to find "solutions" that exacerbate a problem. Qualitative leak testing methods eliminate some of the guesswork.

Christine Beall, AIA, CCS

SEALANT JOINT DESIGN

REFERENCE: Beall, C., "Sealant Joint Design", <u>Water in Exterior Building Walls: Problems and Solutions</u>, ASTM STP 1107, Thomas A. Schwartz, editor, American Society for Testing and Materials, Philadelphia, 1991.

ABSTRACT: Sealant joint failures usually result from either inadequate provisions for movement, improper selection and use of materials, poor workmanship, poor detailing, or combinations of these problems. The proper design of sealant joints includes not only movement calculations and material selection, but planning and detailing to provide overall moisture control, and specifying to assure proper installation. This paper discusses three common joint types, and outlines the primary variables involved in the design of moving joints. Tables of thermal and moisture expansion coefficients are given, and various formulas used for calculating movement are discussed, along with differential movement potential and construction tolerances. An analysis of generic sealant types and characteristics is given, with an emphasis on substrate compatibility, sealant performance and selection. The sum of the information represents a set of criteria which will assist practicing architects and engineers in the joint design process.

KEYWORDS: sealants, construction tolerances, silicones, urethanes, polysulfides, control joint, expansion joint

Sealant joint failures can be catastrophic. The intrusion of water through the exterior envelope of a building can not only deteriorate the envelope materials themselves, but can damage costly interior finishes, equipment and contents. Millions of dollars annually in water damage and construction litigation results when such joints fail due to inadequate provisions for movement, improper selection and use of materials, poor workmanship, poor detailing, or combinations thereof.

The weather integrity of building walls, as they are commonly designed today, relies too heavily on the performance of sealant joints as a single line of defense against

---

Architectural Consultant, Christine Beall • Architect, 5415 Woodview Avenue, P. O. Box 5730, Austin, Texas 78763.

moisture intrusion. Effective moisture control is achieved when design is based not only on the function of individual details, but on the overall performance of the building envelope as a weather barrier, which is formed only in part by sealant joints.

JOINT TYPES

The first step in a comprehensive joint sealing system is selection of a joint type. Butt joints, lap joints and fillet joints are each appropriate to different types of materials and different field conditions. Butt joints are the most common type of site constructed sealant joint. They are used between adjacent, parallel panels that are in the same plane, and may be hourglass shaped, rectangular, or square (fig. 1). Butt joints are typically used as expansion joints, control joints, and perimeter joints between similar or dissimilar components.

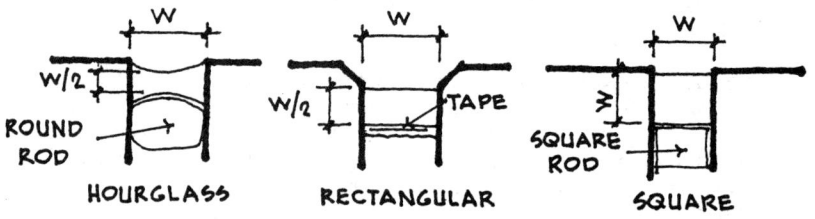

Fig. 1 -- Butt joints

Triangular fillet joints may be used between adjacent panels in non-parallel planes such as the inside corners of walls, or at window and door setbacks where butt joints are either not practical or not feasible (fig. 2).

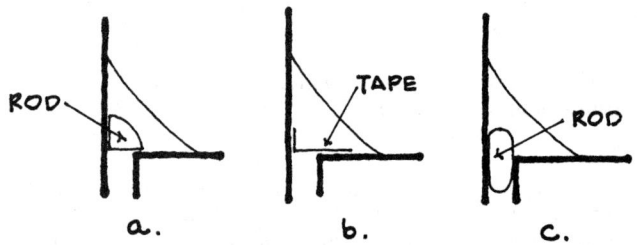

Fig. 2 -- Fillet joints

Lap joints are frequently used in glazing systems, sheet metal fabrications, and external corner joints (fig. 3). Both preformed tapes and liquid sealants are used, sometimes in combination with one another to form primary and secondary seals.

Fig. 3 -- Lap joints

JOINT SIZING

Building envelopes are not monolithic. They are made up of separate components isolated from one another and from dissimilar materials by joints. Buildings, walls, and the materials which they comprise are also in motion, expanding and contracting in response to temperature and moisture fluctuations, and moving in response to applied loads. The joints which separate building components, and the sealants used to fill them, must be able to accommodate such movements, or failure is inevitable. O'Connor [1] and Grimm [2] have developed thorough movement analysis methods, with formulas and coefficients for predicting various kinds of movements. O'Connor's method is cited here.

Thermal Movement

Thermal expansion and contraction is calculated based on a coefficient of thermal movement (C_t), and the difference in substrate surface temperature extremes under in-service conditions. Surface temperatures represent the greatest potential movement and are more accurate predictors than ambient temperatures. Winter surface temperatures (T_w) are established from ASHRAE winter design dry bulb air temperatures [3] since the wall surface in winter will generally be within a few degrees of ambient. Summer wall surface temperatures, however, are affected by both ambient temperature and solar radiation, so radiative heat gain, thermal mass and the temperature gradient through the thickness of the material must be considered. When two dissimilar materials abut the joint, the relative mass of the components should also be considered. O'Connor [1] uses the equation

$$T_s = T_a + XS \qquad (equation\ 1)$$

where:
- T_s = extreme summer wall surface temperature,
- T_a = extreme summer air temperature (dry bulb),
- X = constant for heat capacity of material (table 1), and
- S = wall material solar absorption coefficient (table 2).

The total wall surface temperature differential (ΔT) is found by subtracting winter surface temperature (T_w) from summer surface temperature (T_s). The thermal movement coefficient, temperature differential, and proposed length of the panel or the spacing between joints (L) are then used in the basic formula for thermal movement given in ASTM C962 Standard Guide for Use of Elastomeric Joint Sealants [4]

$$M_t = (C_t)\ (\Delta T)\ (L) \qquad (equation\ 2)$$

where:
- M_t = thermal movement,
- C_t = coefficient of thermal movement (table 3),
- ΔT = total surface temperature differential in °C (°F), and
- L = panel length or joint spacing, in mm (inches).

For example, thermal movement for a brick veneer wall with a thermal expansion coefficient of 6.5×10^{-6} (3.6×10^{-6}), an estimated temperature differential of 65°C (150°F), and a joint spacing of 6 meters (20 ft) would be calculated as

$M_t = (0.0000065)\ (65°C)\ (6000\ mm) = 3\ mm$
$M_t = (0.0000036)\ (150°F)\ (240\ in) = 0.1296\ in$ (or about 1/8 in).

The limited movement capability of the proposed sealant material must be taken into account when sizing joints to accommodate the alternating thermal expansion and contraction just calculated. An elastomeric sealant rated as ±25%, for example,

Table 1 -- Constant for heat capacity of material (X)

	metric[1]	(inch-pounds)	
X =	56	(100)	low heat capacity materials
or			
X =	72	(130)	solar radiation reflected on low heat capacity materials
X =	42	(75)	high heat capacity materials
or			
X =	56	(100)	solar radiation reflected on high heat capacity materials

Notes: 1. Metric constant is derived proportionally from inch-pound constant based on temperature relationship.
2. Materials such as exterior insulation and finish systems and well insulated metal panel curtainwalls have low thermal storage capacity and therefore low heat capacity. Materials such as precast panels and masonry walls, on the other hand, have high thermal storage capacity and therefore high heat capacity.
3. If the wall surface receives reflected as well as direct solar radiation, use the larger constant. Reflected radiation can come from adjacent wall surfaces, roofs, and paving.

Source: O'Connor [1]

Table 2 -- Solar absorptivity coefficients (S)

Material	Coefficient
aluminum, clear finish	0.60
aluminum paint	0.40
mineral board, natural color	0.75
mineral board, white	0.61
brick, light buff color (yellow)	0.50-0.70
brick, red	0.65-0.85
brick, white	0.25-0.50
concrete, natural	0.65
copper, tarnished	0.80
copper, patina	0.65
galvanized steel, unfinished	0.90
galvanized steel, white	0.26
glass, clear (1/4")	0.15
glass, tinted (1/4")	0.48-0.53
glass, reflective (1/4")	0.60-0.83
marble, white	0.58
surface color, black	0.95
surface color, dark grey	0.80
surface color, light grey	0.65
surface color, white	0.45
tinned surface	0.05
wood, smooth	0.78

Source: O'Connor [1]

Table 3 -- Thermal expansion coefficients (C_i)
(multiply by 10^{-6})

Material	Celsius mm/mm/°C	Fahrenheit in/in/°F
Brick		
clay or shale	6.5	3.6
fire clay	4.5	2.5
Concrete Masonry		
normal weight		
sand & gravel aggregate	9.4	5.2
crushed stone aggregate	9.4	5.2
medium weight		
air-cooled slag	8.3	4.6
lightweight		
coal cinders	5.6	3.1
expanded slag	8.3	4.6
expanded shale	7.7	4.3
pumice	7.4	4.1
Stone		
granite	5.0-11.0	2.8 - 6.1
limestone	4.0-12.0	2.2 - 6.7
marble	6.7-22.1	3.7 -12.3
sandstone	8.0-12.0	4.4 - 6.7
slate	8.0-10.0	4.4 - 5.6
travertine	6.0-10.0	3.3 - 5.6
Concrete		
calcareous aggregate	9.0	5.0
siliceous aggregate	10.8	6.0
quartzite aggregate	12.6	7.0
Metals		
aluminum	23.8	13.2
steel, carbon	12.1	6.7
steel, stainless		
301 alloy	16.9	9.4
302/304 alloy	17.3	9.6
410 alloy	11.0	6.1
brass	18.7	10.4
bronze	18.0-20.9	10.0 - 11.6
copper	16.9	9.4
iron		
cast	10.6	5.9
wrought	12.1	6.7
lead	28.6	15.9
Glass	8.8	4.9
Plastic		
acrylic sheet	74.0	41.0
polycarbonate sheet	68.4	38.0

Note: Coefficients taken from ASTM C962 Standard Guide for Use of Elastomeric Sealants [4] and other sources. Refer to ASTM C962 for a more complete listing of materials.

reportedly can tolerate a maximum extension of +25% of the joint width, and maximum compression of -25% of the joint width (when tested under laboratory controlled conditions, in accordance with ASTM C719 [5]). In other words, the fully compressed sealant will occupy three fourths of the designed joint width, leaving only one fourth of the width for actual unrestrained movement. To account for this, joints must be sized four times the calculated movement for ±25% sealants, and two times the calculated movement for ±50% sealants. For example four times a calculated thermal movement of 3 mm (1/8 in) would require a joint 12 mm (1/2 in) wide. When the joint opens 3mm (1/8 in) the sealant is extended to its full +25%, and when the joint closes 3 mm (1/8 in), the sealant is compressed the full -25%. A sealant claimed as +100/-50% is governed by its compressibility, so the joint still must be twice the calculated movement. ASTM C962 [4] accommodates sealant movement capability using the formula

$$J_t = (100/S_m) (M_t) \qquad \text{(equation 3)}$$

where: J_t = joint width required for thermal movement
S_m = sealant movement capacity, and
M_t = calculated thermal movement.

Using the movement calculated in equation 2 above, and a sealant rated at ±50%,

$$J_t = (100/50) (3 \text{ mm}) = 6 \text{ mm}$$
$$J_t = (100/50) (1/8 \text{ in}) = 1/4 \text{ inch.}$$

O'Connor, however, recommends that sealants be used at a percentage of their rated movement capacity to allow for imprecisions in establishing surface temperatures, imperfect workmanship (i.e., non laboratory controlled conditions), and other unknowns. The amount of reduction should depend on the particular circumstances of a joint design and the factor of safety desired. With this additional limitation, using the sealant at only 80% of its rated capacity for this design, the formula then would be

$$J_t = (100/0.8S_m) (M_t) \qquad \text{(equation 4)}$$

$$= [100/(0.8)(50)] (3 \text{ mm}) = 8 \text{ mm}$$
$$= [100/(0.8)(50)] (1/8 \text{ in}) = 5/16 \text{ in.}$$

Moisture Movement

For most materials, calculation of thermal movement alone for joint design is not enough. Concrete, brick and natural stone also experience dimensional changes with changes in moisture content. Reversible moisture movement is based on the likely extremes of in-service moisture content, which will decrease as the wall surface temperature rises, and increase as the wall surface temperature falls. Such reversible moisture contraction and expansion, however, is offset to some degree by opposing thermal expansion and contraction. Even though the compensating movements do not necessarily occur simultaneously, the overall effect of reversible moisture movement is generally considered negligible.

Brick, concrete, and concrete products also experience irreversible moisture movements that are cumulative with thermal movement. Irreversible moisture movement is based, not on in-service conditions, but on the age of the material. Even though the net effect of thermal and moisture movement may be difficult or impossible to determine precisely, we do know two things: (a) brick expands permanently and irreversibly with exposure to repeated freeze/thaw cycles [2] and to atmospheric moisture; and (b) concrete, concrete masonry and stucco all experience permanent and irreversible shrinkage during the curing process.

Expansion joints must be used in brick to absorb expansive movements (fig. 4), and control joints must be used to control the location of shrinkage cracking in concrete materials (fig. 5).

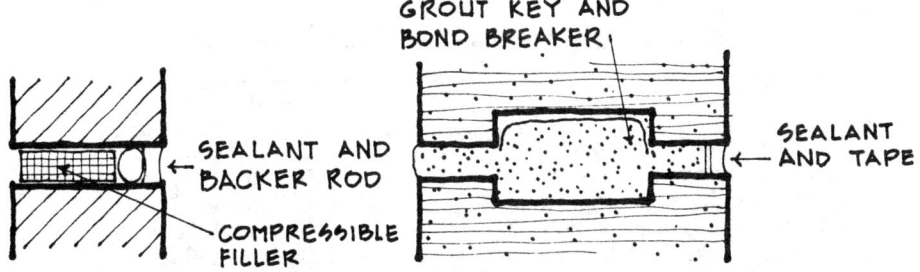

Fig. 4 -- Brick expansion joint   Fig. 5 -- Concrete masonry control joint

These two terms are not interchangeable. Brick expansion joints or soft joints (i.e., without mortar), even while they are experiencing cyclical expansion and contraction due to temperature changes, are also experiencing a permanent closure caused by permanent moisture expansion of the units. Control joints in concrete, concrete masonry and stucco, on the other hand, are weakened planes designed to localize the cracking that accompanies material shrinkage. Concrete masonry control joints can contain mortar because the initial material shrinkage leaves the joint open to accommodate subsequent thermal expansion. A grout column is used as a shear key, and the face of a control joint is raked out and sealed to prevent moisture intrusion. An expansion joint is formed with backer rod and sealant in an open void, or in front of a compressible pad that is used to keep mortar out during construction. An expansion joint that is bridged by mortar extrusions from the joint cannot accommodate the required movement.

The absence or infrequency of functioning expansion and control joints in brick and concrete masonry respectively is a common cause of cracking and moisture leakage. Irreversible moisture expansion and contraction in concrete and masonry construction must be added to thermal movement in sizing sealant joints. O'Connor lists a range of coefficients for irreversible moisture-induced movement for several brick and concrete products (table 4), and uses a formula for calculating movement

$$J_m = (C_m/100)(L) \qquad \text{(equation 5)}$$

where
J_m = joint width for moisture movement,
C_m = coefficient of moisture movement (table 4), expressed as a decimal, and
L = panel length or joint spacing.

Using a brick with a coefficient of +0.05, and a panel length of 6 meters (20 ft), the joint width required for moisture movement would be

$J_m = (+0.05/100)(6000) = +3$ mm expansion
$J_m = (+0.05/100)(240) = +0.12$ (or about 1/8 in) expansion.

When added to the width calculated for thermal movement above (equation 4), the joint width required increases to

$J_t + J_m$ = 8 mm + 3 mm = 11 mm
     = 5/16 in + 1/8 in = 7/16 in.

Because of the effects of wall length/height ratios and the spacing of joint reinforcement or bond beams, the National Concrete Masonry Association (NCMA) uses empirical recommendations rather than calculations for control joint spacing (table 5).

Table 4 -- Moisture movement coefficients (C_m)

material	movement (%)
concrete, gravel aggregate	-0.03 to -0.08
concrete, limestone aggregate	-0.03 to -0.04
concrete, lightweight aggregate	-0.03 to -0.09
concrete block, dense aggregate	-0.02 to -0.06
concrete block, lightweight aggregate	-0.02 to -0.06
face brick, clay	+0.02 to +0.07

Notes: 1. + indicates permanent material expansion, - indicates permanent material shrinkage.
2. If specific data for a particular unit are not available, use the maximum value given in the table.

Source: O'Connor [1]

Table 5 -- Control joint spacing for moisture-controlled, Type I CMU

	vertical spacing of joint reinforcement			
	none (none)	60 cm (24 in)	40 cm (16 in)	20 cm (8 in)
expressed as ratio of wall length to wall height (L/H)	2.0	2.5	3.0	4.0
control joint spacing not to exceed	12 m (40 ft)	14 m (45 ft)	15 m (50 ft)	18 m (60 ft)

Note: With reinforced bond beams 1.2 meters (4 ft) on center vertically in lieu of joint reinforcement, control joints may be spaced a maximum of 18 meters (60 ft) on center.

Source: NCMA [6]

The Brick Institute of America (BIA) uses an average coefficient of moisture expansion for all brick of 0.05%, which is added to thermal movement in a slightly different equation [7]

$$E = [0.0005 + (C_t)(\Delta T)](L) \qquad (equation\ 6)$$

where
  E = expansion of brick wall,
  C_t = coefficient of thermal expansion,
  ΔT = temperature differential, and
  L = panel length or joint spacing.

Using the same material, coefficient and panel length as in the previous examples, the wall expansion calculated by this method would be

$$E = [0.0005 + (0.0000065)(65)](6000) = 6 \text{ mm}$$
$$E = [0.0005 + (0.0000036)(150)](240) = 1/4 \text{ in.}$$

For a sealant rated ±50%, twice the calculated movement gives a 12 mm (1/2 in) required joint width, or approximately the same as O'Connor's method. Table 6 summarizes basic brick expansion joint spacing for various standard joint widths up to 24 mm (1 in). For both brick and concrete masonry, additional vertical expansion or control joints must be located at openings, pilasters, offsets, returns, intersecting walls, and at changes in wall height or cross section.

Table 6 -- Joint width and spacing for brick walls
65°C ΔT (150°F ΔT)

anticipated expansion (W)	minimum joint width		maximum joint spacing (L)
	(2 x W) ±50% S_m	(4 x W) ±25% S_m	
3 mm (1/8 in)	6 mm (1/4 in)	12 mm (1/2 in)	3.5 m (12'-0")
5 mm (3/16 in)	10 mm (3/8 in)	20 mm (3/4 in)	5.5 m (18'-0")
6 mm (1/4 in)	12 mm (1/2 in)	24 mm (1 in)	6 m (20'-0")

Horizontal Joints

In addition to thermal and moisture movements, horizontal expansion joints in veneers and curtainwalls also experience permanent contractions caused by load deflections and frame shortening. Beams can be cambered and formwork adjusted slightly in height, but a certain amount of vertical movement is unavoidable.

Deflection of beams and shelf angles from live load and long term creep can be calculated by the structural engineer. The residual frame shortening that takes place after the cladding is installed depends on the time lapse after erection of the frame. Grimm [2] has calculated that some concrete frames may lose as much as 3 to 5 mm (1/8 to 1/4 inch) in 3.5 meters (12 ft).

Both brick veneer and metal curtainwall require particular care in calculating the size of the expansion joints or soft joints that are required below all shelf angles. At the same time the frame is shortening, metal curtainwall sections experience large thermal movements, and brick undergoes permanent vertical expansion. Inadequate allowance for these opposing movements is a common cause, not only of sealant joint failure, but physical damage to the cladding as panel sections are literally squeezed between supports. The resulting moisture penetration causes additional structural damage as metal anchors, steel backing studs and the shelf angles themselves begin to corrode.

CONSTRUCTION TOLERANCES

Materiala, fabrication and erection tolerances must also be considered in sizing sealant joints. When unanticipated construction tolerances increase or decrease the designed joint width, the expected thermal and moisture movement remains the same, but sealant performance is dramatically affected. Joints that are slightly larger than necessary do not present as much of a problem as joints that are too small.

If a 12 mm (1/2 in) joint is designed to accommodate 3 mm (1/8 in) movement in both compression and extension, this represents a 25% strain. With a ±6 mm (±1/4 in) construction tolerance, however, the as-built joints could be as wide as 18 mm (3/4 in) or as narrow as 6 mm (1/4 in). In the larger joint, the anticipated movement will create only a 16% strain in the sealant. But in the smaller joint, a ±3 mm (±1/8 in) movement generates a 50% strain (fig. 6). Only the highest performance sealants can tolerate that much movement. Anything less will fail, and the joint will leak within one seasonal cycle [8].

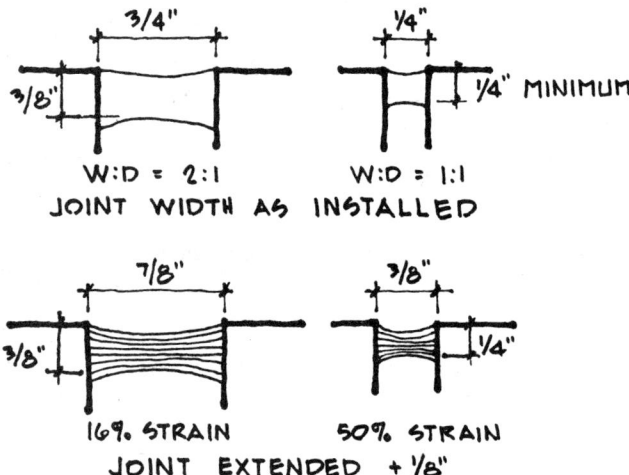

Fig. 6 -- Sealant strain

Since many sealant failures are related to undersized joints, allowance for construction tolerances could make a critical difference in preventing water leaks. Construction tolerances are set within each industry, based simply on judgement and economic considerations [9]. That is, each industry develops its own standards based on its own collective judgement and its own economic consideration of what is reasonable and cost effective. A lack of coordination among various systems and components results in inevitable problems of misfit that are taken up in the joints. Some allowance must be made in designing the joint width and/or determining joint spacing to permit what the architect or engineer determines to be a "reasonable" combined or net construction tolerance for the materials and systems involved.

To determine the total joint width required, construction tolerances are added to the sum of the joint widths ($J_t + J_m$) required for thermal and moisture movements. If the width of the joint is aesthetically unacceptable, joints can be spaced more frequently to reduce the required width.

Once the joint is sized, it must be properly detailed so that sealant application is practical, the backer rod can be placed and held in compression, and the joint sides provide adequate adhesion surfaces.

JOINT PERFORMANCE

Installation temperatures have an effect on the way sealant joints perform. When sealants are installed at moderate temperatures, cold weather causes sealant extension and warm weather causes compression. If installed at low temperatures, however, the sealant is in compression in all but extremely cold conditions, and permanent deformation or extrusion from the joint is possible. Irreversible moisture expansion in brick masonry can add to the effect. If sealants are installed when surface temperatures are extremely high, they are in extension most of the time. Residual moisture shrinkage in concrete and concrete masonry, combined with setting or curing shrinkage in the sealant itself impose additional stress, not only on the sealant, but also on the substrate if a high modulus sealant is used. Tensile failure in adhesion or cohesion of both the sealant and the substrate often result.

Figure 7 illustrates the dynamics of joint movement and sealant performance at various installation temperatures [10]. The closer the installation temperature is to the mean, the less strain the sealant will undergo in service and the better it will perform. Although the installation temperature ranges recommended by sealant manufacturers often have more to do with application problems than cyclical joint movements, moderate installation temperatures will generally result in better sealant performance.

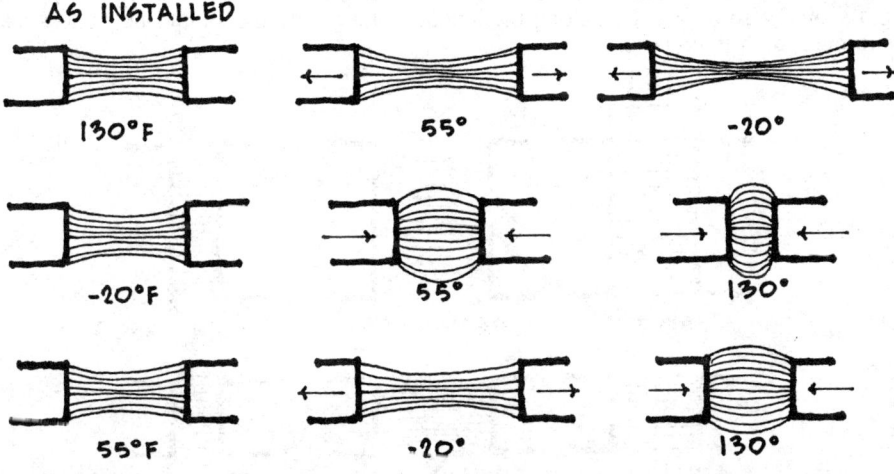

Fig. 7 -- Installation temperatures

The inherent flexibility of some cladding systems will permit the adjustment of minimum joint size at the time of installation [11]. Specifications should call for a minimum joint width for conditions of both hot weather and cold weather installation. Field adjustments then, can allow better performance of the sealant joints through full seasonal temperature cycles.

Movement of the joint during sealant cure can also cause failure. Compression set results when a joint closes in hot weather before full cure is effected. When the joint re-opens, the sealant experiences high internal and adhesive stresses that may cause rupture or spall the face of the joint (fig. 8). Expansion set occurs when a joint opens in cold weather and stretches the sealant before it is cured. Re-closure causes the material to bulge out of the joint. Compression set failures are more common because heat accelerates the chemical reactions. (The term compression set is also used to refer to permanent compression in fully cured elastomers.)

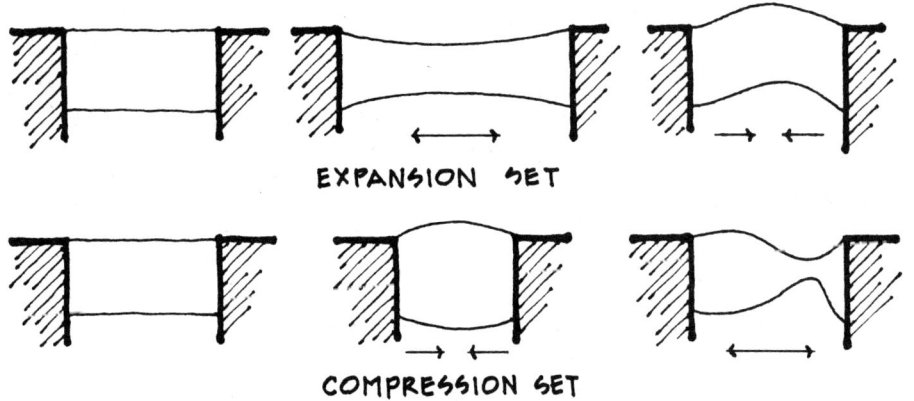

Fig. 8 -- Movement during cure

For sealants to function properly, joint shape or geometry must also be considered. The depth of the sealant should generally be less than or equal to the width. For most butt joint applications, a 2:1 width to depth ratio is recommended. Increasing the depth increases the strain in the sealant as well as at the bond line (fig. 9). Sealant depth should also be constant along the length of the joint, and should be no less than a minimum of 6 mm (1/4 in).

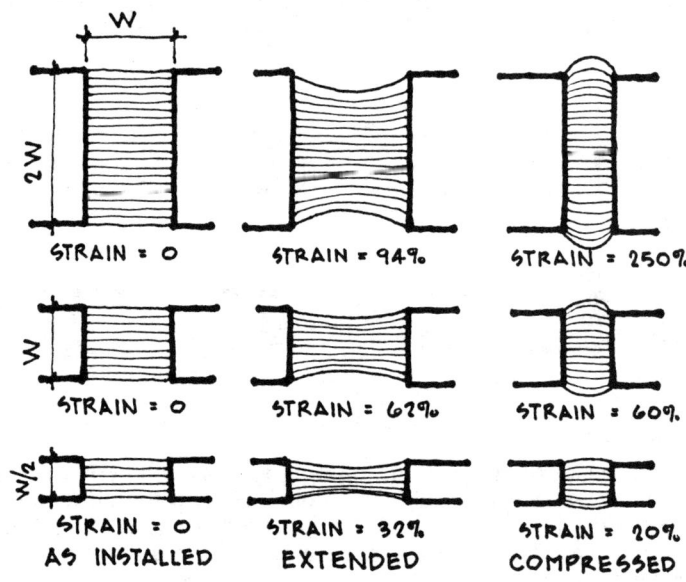

Fig. 9 -- Sealant configuration (adapted from ref. 12)

In lap joints, shear stresses govern sealant behavior, and the magnitude of the stress is related both to the movement and to the sealant thickness. For installations made at moderate temperatures, the thickness of the sealant should generally be at least one half the anticipated movement, with a minimum thickness of 6 mm (1/4 in). Where installation will occur at higher or lower temperatures, the sealant thickness should equal the anticipated movement, again with a minimum thickness of 6 mm (1/4 in) [12].

Myers [13] has investigated the performance of fillet joints, and recommends a triangular or quarter round backer rod for best theoretical sealant configuration. However, these materials are not readily available, so the use of bond breaker tape and the adaptation of conventional round backer rods (see fig. 2) is more common. The width of the bond breaker can be calculated by dividing the combined thermal and moisture movement by the sealant rating [13]. If the joint is expected to move ±5 mm (3/16 in) and the sealant is classified as ±25%, the bond breaker or release tape (W) must be 20 mm (3/4 in) wide (fig. 10). The adhesion width (B) should be a minimum of 10 mm (3/8 in), and the sealant thickness (T) a minimum of 6 mm (1/4 in).

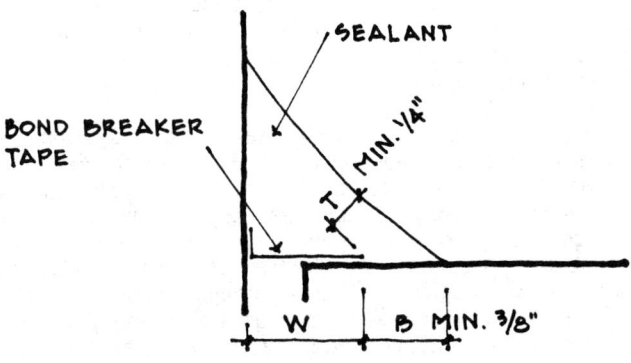

Fig. 10 -- Fillet joint shape [13]

SEALANT MATERIALS

Each type of sealant has strengths and weaknesses which must be considered based on the requirements of a particular application. The following is a general discussion of sealant properties. Each individual sealant formulation, however, will vary from one proprietary product to the next, even within a generic type. When choosing a caulk or sealant, always request test data from the manufacturer to verify reported properties, including information on long term durability.

Polysulfides, in general, have a reported movement capability of ±25%. Adhesion to steel, aluminum and glass is very good, and generally does not require priming. Porous substrates such as marble, limestone, granite, brick, concrete, concrete block, wood and plastics must be primed to achieve serviceable adhesion. Primers for polysulfides are made and furnished by the sealant manufacturer, developed for use with their own specific sealant formulation. These primers are not interchangeable among different proprietary sealants.

Two-part polysulfides have good extensibility and good resistance to weathering and aging. At 75%, recovery from extension and compression is less than that of urethanes and silicones. Special formulations can provide excellent solvent resistance to many chemicals and petroleum products such as gasoline and jet fuel. Tack-free time is 36-48 hours, which will invite some dirt pickup. Full cure generally takes a week in good weather, longer in cooler temperatures.

One-part polysulfides have similar properties except that recovery is poorer and cure rate is much slower. Cure rate depends on ambient moisture and oxygen. With moderate temperatures and high humidity, tack-free time is reported by manufacturers to be as long as three days. Full cure may take 30-45 days under optimum conditions and much longer in cold, dry weather. After the sealant has been properly tooled into the joint, a light water spray can speed initial skinning and help prevent excessive dirt pickup. Wet tooling, however, is not recommended because of possible contamination of the joint surfaces and other problems. Because of the slow curing time, both one and two-part polysulfides are susceptible to compression set and extension set. Polysulfides do experience hardening with age (known as heat aging). As the hardness of a sealant increases, so does its modulus of elasticity, with an attendant increase in the tendency for adhesion failure.

Urethane sealants are available in both one-part and two-part systems. Reported movement range is ±25%, ±50% and +100/-50% on different formulations. To a great extent, urethanes have replaced polysulfides in the United States. A wide variety of performance characteristics are available.

Urethanes have good extensibility, excellent resistance to abrasion, paintability, long service life and good adhesion to most substrates, especially concrete, masonry and metals. Resistance to ultraviolet and ozone deterioration is only fair, with most urethanes having a tendency to craze. Primers are required, however, for concrete contaminated with form oils or other substances, and urethanes generally are not recommended for glazing applications. At 80-90%, recovery is much better than polysulfides, but not as good as silicones. The two-component sealants are extremely well suited for horizontal traffic joints. Some urethanes also have good chemical and solvent resistance. Shrinkage is negligible, and tear resistance is excellent if the sealant depth is sufficient. Urethane sealants can experience some hardening over time.

Some light-colored urethanes will discolor with time, but dirt pickup is a problem only with the one-part systems. Tack-free time is longer than for two-part materials (12-36 hours in good weather, several days in cold, dry weather), and full cure takes at least 18-21 days. Hot, humid conditions will produce a faster cure. If urethanes do not pick up dirt during cure, they will stay clean for many years. This results from a so-called "self-cleaning" characteristic. Surface oxidation causes the sealant to chalk and continuously slough off any accumulated dirt. It is rare, however, for urethanes to attract and hold airborne particles.

All urethanes are extremely sensitive to moisture before and during cure. A damp substrate will cause the formation of bubbles at the bond line which prevent effective adhesion. Concrete should be fully dry to the touch for at least 24 hours before urethane sealants are applied. After a rain, allow 2-3 days dry-out time before sealing joints. Urethanes are more forgiving than silicones in regard to joint cleaning, but surface laitance and loose particles must be removed to get good adhesion.

Silicone sealants come in a variety of formulations that include one-part, two-part and three-part mixes. The multi-component materials generally have a very rapid cure rate of 15 minutes to 3 hours, but tooling time is also shorter. Silicones are gunnable even in very cold weather, and have excellent extension and movement in the cold. They also retain their strength and shape in hot weather, and the opaque sealants are highly resistant to ultraviolet and ozone deterioration. Once cured, silicones experience very little hardening over time.

The reported movement ability of silicone sealants will vary with the formulations. Some high-modulus products fall in the low range of ±12.5%, others in the mid-range of ±25%. High performance, low-modulus silicones may tolerate ±50% movement or as much as 100% extension and 50% compression (+100/-50%).

Recovery with high-modulus silicones is 85-99%. The low-modulus materials have 70-90% recovery. Silicones are chemically stable and extremely resistant to weathering and aging. Silicones have unprimed adhesion to most substrates, but surface cleaning is critical. Tenacious bond to glass makes silicones the only sealants permitted for structural glazing. Anodized or coated aluminum, however, must be stripped or primed first because most silicones will not adhere to the slick surface contaminants deposited during manufacture. Some silicones, however, are formulated for unprimed adhesion to anodized aluminum. Silicones abrade more easily than urethanes, so horizontal deck, plaza, sidewalk, or other traffic-bearing joints should be recessed sufficiently to avoid abrasion wear.

Dirt pickup may be one of the most widely voiced complaints about silicone sealants. Fast skinning characteristics prevent silicones from imbedding blowing dust at the job site into an uncured surface. Static charges on the surface of cured silicones, however, do attract dirt (more so than either urethanes or polysulfides), especially the softer, low-modulus formulations. Rain can wash this surface dirt down on to adjacent materials, leaving noticeable stains. Rain cannot completely remove this dirt buildup, however, so it must be scrubbed off during regular maintenance. The plasticizers in some silicones will also stain porous substrates, so pre-construction testing is recommended.

Sealant selection must always be based on service conditions and substrate. If a specific proprietary product is not selected, specify by generic type and compliance with proper Type, Class and Use designations of ASTM C920 Standard Specification for Elastomeric Joint Sealants [14]. Project specifications should also require that sealants be installed only by trained applicators.

SUBSTRATE COMPATIBILITY

Some materials offer very poor substrates for almost any caulk or sealant. Wet, frozen, contaminated, deteriorated, weak, unsound and unstable surfaces will prevent adhesion. Adhesion is also affected by concrete form release compounds, hot-dip galvanizing, polyethylene coatings on spandrel panels, self-oxidizing steel, mill-finish aluminum, high-performance glass coatings, and fluoropolymer metal coatings that are designed to resist soiling and weathering.

Substrate compatibility is also a potential problem. Some formulations will stain porous substrates such as natural stone. Acetoxy silicones (those that release acetic acid) should not be used on marble, galvanized steel, copper, concrete or other cementitious materials, or on any surface that is prone to attack or corrosion by weak acids. Silicones that release amines or oximes should not be used on copper [10].

Exterior insulation and finish systems (EIFS) present another kind of problem. Polystyrene insulation board is a very weak substrate. Unprotected EPS board is easily ripped away by high-modulus sealants in extension. To prevent such tearing, the insulation board edges must be wrapped with mesh and base coat, or covered by edge and joint accessories such as J-molds. The EIFS finish coat should not be returned into the joint because it can easily soften and delaminate if moisture is present behind the finish, causing joint failure.

Glass fiber reinforced concrete (GFRC) panels are also problematic. The material responds so rapidly to temperature changes that compression and extension set often occur. The only way to compensate for this with gunnable sealants is to use a low-modulus material with a fast cure. Accessories in glazing systems and materials used for backer rods and bond breaker tapes must also be compatible with the sealant. Solvent cleaners and primers can prevent adhesion if they are not compatible or are improperly applied, and some sealants react with the polyvinyl butyral (PVB) layer in laminated glass, causing localized delamination or discoloration of the PVB.

SUMMARY

The joint design process has five components: (1) movement calculation; (2) joint sizing; (3) proper detailing; (4) product analysis and selection; and (5) proper specifications. Only a small effort is required to build in a margin of safety which can significantly reduce the ultimate cost of maintenance and repair and the incidence of moisture damage. Ultimately, architects and engineers must move away from the use of sealant joints as the first and only line of defense against the weather, and toward detailing that provides protected joints and natural moisture control.

REFERENCES

1. O'Connor, Thomas F., "Design of Sealant Joints", Building Sealants: Materials, Properties and Performance, ASTM STP 1069, Thomas F. O'Connor, editor, Philadelphia: American Society for Testing and Materials, 1990, pp. 141-164.

2. Grimm, Clayford T., "Probablistic Design of Expansion Joints in Brick Cladding", Proceedings of the Fourth Canadian Masonry Symposium, Fredericton, New Brunswick, Canada, June 1986, pp. 553-568.

3. American Society of Heating, Refrigerating and Air Conditioning Engineers, Handbook of Fundamentals, ASHRAE, Atlanta, 1989.

4. "ASTM C962, Standard Guide for Use of Elastomeric Joint Sealants", Annual Book of ASTM Standards, Volume 04.07, American Society for Testing and Materials, Philadelphia, 1989.

5. "ASTM C 719, Standard Test Method for Adhesion and Cohesion of Elastomeric Joint Sealants Under Cyclic Movement (Hockman Cycle)", Annual Book of ASTM Standards, Volume 04.07, American Society for Testing and Materials, Philadelphia, 1989.

6. National Concrete Masonry Association, "TEK 53", Herndon, Virginia, 1973.

7. Brick Institute of America, "Tech Note 18A, Revised", Reston, Virginia, 1988.

8. Panek, Julian R. and John P. Cook, Construction Sealants and Adhesives, 2nd edition, New York: John Wiley & Sons, 1984.

9. Beall, Christine, "Construction Tolerances", The Construction Specifier, Vol. 43, No. 6, August 1990, pp. 74-49.

10. Klosowski, Jerome M., Sealants in Construction, New York: Marcel Dekker, Inc., 1989.

11. Beall, Christine, "Joint Design and Sealant Selection in Masonry Construction", Proceedings of the Fifth Canadian Masonry Symposium, Vancouver, British Columbia, June 1989, pp. 487-496.

12. American Concrete Institute, "Guide to Joint Sealants for Concrete Construction", ACI, Detroit, 1977.

13. Myers, James C., "Sealant Configurations and Performance", Architectural Record, Vol. 178, No. 1, January 1990, pp. 150-151.

14. "ASTM C920, Standard Specification for Elastomeric Joint Sealants", Annual Book of ASTM Standards, Volume 04.07, American Society for Testing and Materials, Philadelphia, 1989.

Andrew Charles Yanoviak, AIA, CSI

WATER INFILTRATION IN HAWAII and ENSUING CONSTRUCTION LITIGATION

REFERENCE: Yanoviak, A. C., "Water Infiltration in Hawaii and Ensuing Construction Litigation," *Water in Exterior Building Walls: Problems and Solutions, ASTM STP 1107*, Thomas A. Schwartz, Ed., American Society for Testing and Materials, Philadelphia, 1991.

ABSTRACT: Hawaii has its share of water infiltration problems and remedial repair solutions that have also resulted in significant failures attributed to improper design and construction of exterior building walls. Hawaii also has its share of design and construction lawsuits and apportioned culpability. It is well known that this can be a very expensive way to learn, as it affects profitability. It has become increasingly apparent that in many instances the building codes and standards are minimal at best.

KEYWORDS: water infiltration, exterior building walls, lawsuits, materials, products, applications, specifications, design details, building standards, seals and sealants, deficiencies, quality control, testing, technical literature.

A presentation of case studies encountered in Honolulu which exemplify serious water infiltration problems involving property damage and product application failures. Included are architectural and structural design detailing problems and defective specifications for combinations of various building materials with different chemical reactions. Also covered are example of various materials and standards being used for building seals and sealants in both window wall and curtain wall exteriors. Noted are deficiencies in manufacturing quality control standards, technical literature data, inapplicable laboratory test results, and wind tunnel studies for highrise building design, product application, installation and construction. Proposed workable and unworkable remedial repair solutions, and material and design application standards are also noted. Technological challenges presented by architects and engineers to manufacturers and contractors are also included. The effects of weathering, staining, climate, environmental pollution, and ultraviolet radiation on premature and accelerated failures in concrete highrise construction for offices and condominiums in Hawaii are also mentioned.

Andrew Charles Yanoviak, AIA, CSI is an architect-planner and construction litigation consultant with Andrew Charles Yanoviak, AIA; 1188 Bishop Street / Suite 3011; Honolulu, Hawaii 96813

Hawaii is currently experiencing its second 'tidal wave' of major resort hotel and related commercial development. Unfortunately, it appears from on-site observations of construction quality, that Hawaii contractors, architects, and engineers did not learn a whole lot from the first wave of major construction where several tower cranes dotted the skyline simultaneously.

In Reference 1, I clearly illustrated the differences between 'profitability' and 'culpability' with respect to private development in Hawaii. As the contrasting diagrams vividly depict, the architect/engineer('A/E') generally assumes 50% culpability in exchange for about 3% of the total compensation; while, the owners and financiers ('O') on the other hand enjoy 50% of the profitability for only 1% culpability. See Figures 1 and 2 below.

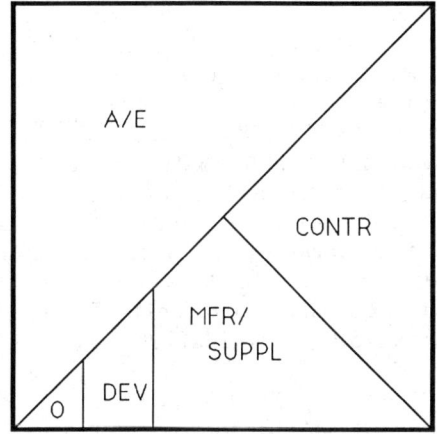

Figure 1: "CULPABILITY"

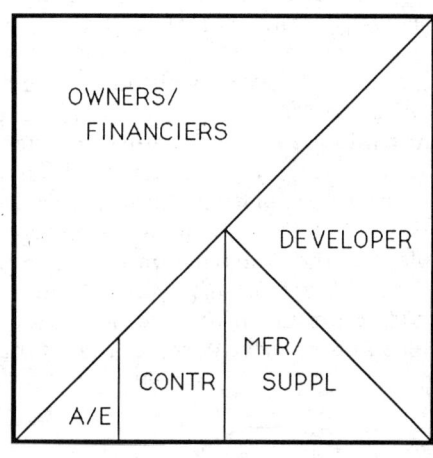

Figure 2: "PROFITABILITY"

The developers ('DEV') also do very well with 25% of the profit and only about 4% of the liability. However, the manufacturers and suppliers ('MFR/SUPPL') only get 12% of the profits for assuming 20% of the culpability. And the contractors ('CONTR') assume 25% of the liability responsibility in exchange for only 10% profitability. Obviously, the costs and benefits of engaging in private speculative development are grossly skewed and imbalanced in favor of the financiers, owners, and developers, to say the least, and this may very well be the root cause of construction deficiencies in Hawaii.

As I have clearly illustrated in Figure 9, there is a definite and compelling need to judiciously balance risk-taking and compensation, as well as liability responsibility and culpability with profitability, among all of the team players involved in private development in the construction industry. These factors should be taken into consideration at the outset, along with design and construction management fees in business negotiations and contractural arrangements, in order to improve the quality of design and construction and mitigate water infiltration problems in private speculative developments.

It is well known among construction litigation consultants, expert witnesses, attorneys, and professional liability insurance carriers, that *'cost-savings'* (or what may be inappropriately called *'value engineering'*) substitutions for products and materials originally specified in accordance with higher quality design standards, are often the sole primary cause for major water infiltration problems and damages as well as ensuing construction litigation in Hawaii. True value engineering should optimize the quality of design and construction, by including many factors from initial development costs to operational maintenance costs.

The outcomes of Figures 1 and 2 are based on the typical traditional triangular contractual relationships between *'Owner'*, *'Contractor'*, and *'A/E'*, where the Owner and the Contractor have a written contract, and the Owner and the A/E have a written agreement; however, there is no direct or even indirect contractual relationship between the A/E and the Contractor as shown in Figure 3. On most private speculative and institutional development projects, there is also a *'Construction Manager'* (*'CM'*) with a direct contractual relationship to the Owner, but no contractual relationships with either the A/E or the Contractor. Figure 4 also shows a contractual role that the Construction Manager fulfills as a representative agent for the Owner, where *'quality control'* and adherence to established *'design standards'* communications are facilitated between the *'Manufacturers*(*'MFR REP'*) *Sales Representatives'* and the *'Manufacturers/Suppliers'* (*'MFR/SUPPL'*).

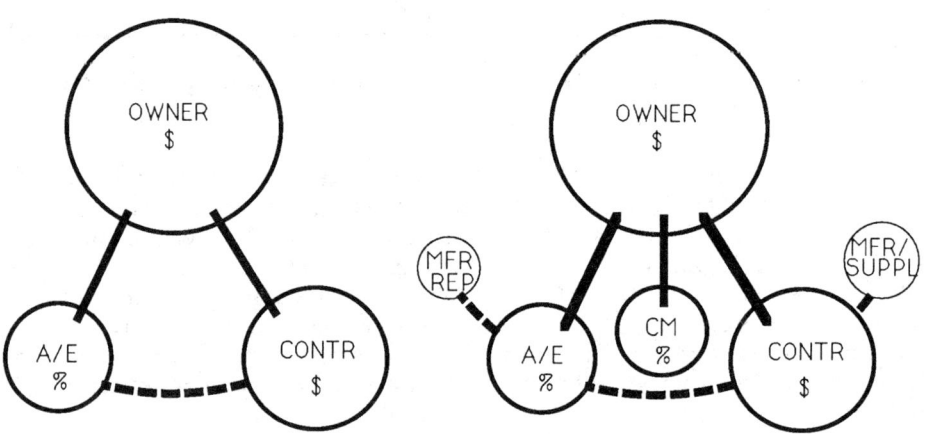

Figures 3 & 4: 'TRIANGULAR CONTRACTUAL RELATIONSHIPS'

Construction Management is a growing area of activity and several large A/E firms have established separate departments or companies to handle the ever increasing workload with the inherent complexities of construction industry practices. Major problems can develop with the contract arrangements depicted in Figure 4, which often require the services of a full time Construction Manager ('CM') to resolve. For example, as a part of his traditional scope-of-work which is normally based on a lump sum or percentage fee, the A/E during the 'design development' stage, generally communicates only with the Manufacturers Sales Representatives in the process of writing the contract specifications. However, the 'job-shopper' Contractor and his 'Developer-Client' (Owner) often communicate directly with factory personnel regarding the manufacture and supply of component building products such as hardware, windows, doors, balcony and stair railings, and elevators, etc. Consequently, what the A/E reviews in the technical literature contained in the Manufacturer's catalog when he makes his selections for the project specifications, and in some instances, even those product components shown and approved on the 'Shop Drawings', may not be entirely recognizable after installation and may have to be rejected if discovered prior to occupancy. If not, they may eventually lead to water infiltration problems or other construction deficiencies or safety hazards and ensuing construction litigation. In those situations, where the Construction Management specialist has also served the Owner/Developer as a so-called 'value engineering' or 'cost-savings' expert, major liability responsibility may be assumed in the process.

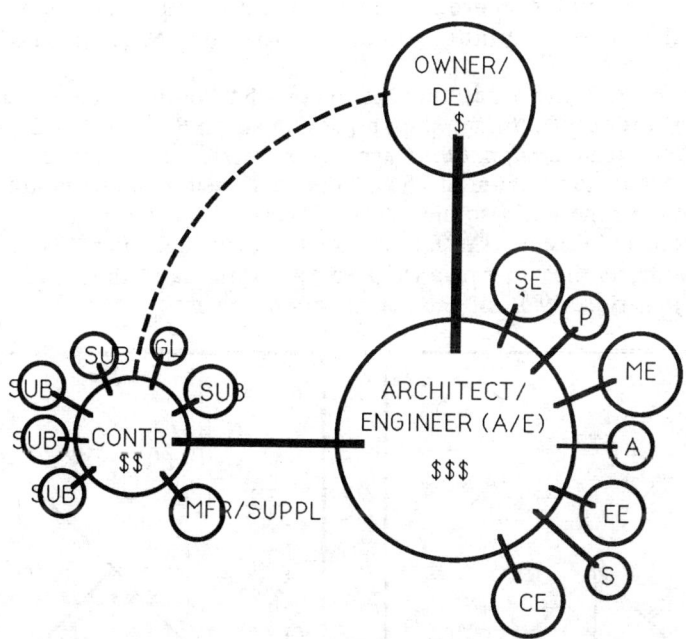

Figure 5: "DESIGN-BUILD CONTRACTURAL RELATIONSHIPS'

The 'Design-Build' concept is becoming more prevalent in conjunction with 'fast-track' construction, where the Owner/Developer no longer contracts directly with the A/E, but rather, retains a Contractor who in turn retains the A/E and keeps his creative spirit well disciplined. The consequence of this particular arrangement is that unfortunately, the frustrated A/E (ARCHITECT/ENGINEER) may have procedurally relinquished many aesthetic and technological design decisions. In the process, since the A/E has lost direct responsible contact with the Owner/ Developer ('DEV'), this often results in overall project 'quality control' problems. Some aggressive and well capitalized A/E's have reacted strongly to this arrangement of contractual relationships by seizing control, as well as maintaining direct and responsible communications with the Owner/ Developer. Realizing that many Contractors ('CONTR') are nothing more than 'brokers' hiring and coordinating Subcontractors ('SUB'), some A/E's have decided to take complete charge of both the 'design' and 'construction' functions, and hopefully improve 'quality control' in the process.

In Figure 5 above, the A/E design team consists of the Structural Engineer ('SE'), Mechanical Engineer ('ME'), Electrical Engineer ('EE'), Civil Engineer ('CE'), Soils Engineer ('S'), Plumbing Engineer ('P'), and Acoustical Consultant ('A'). The Glazing Subcontractor ('GL') is singled out from the array of Subcontractors in the diagram, because of the predominant number of water infiltration and ensuing construction litigation problems in Hawaii and elsewhere involving exterior wall performance. In Reference 2, I reported on several case studies where serious water infiltration had been attributed to the installation of aluminum windows in the midwestern part of the U.S.

Figures 6 & 7 comparatively illustrate what happens to the factors of 'culpability' and 'profitability' shown previously in Figures 1 & 2, resulting from this advanced 'Design-Build' arrangement shown in Figure 5. Notice that the A/E has now assumed 75% of the responsibility instead of 50% as before; however, he has also increased his 'profitability' from 3% to a full 25%. Therefore, individual A/E firms may decide their own 'risk management' options, as they increase their compensation and their overall professional 'quality control' of the construction project.

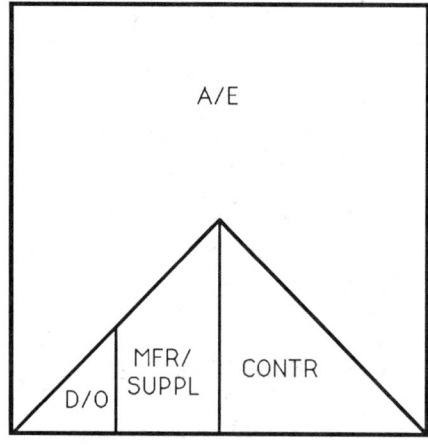

Figure 6: "CULPABILITY"

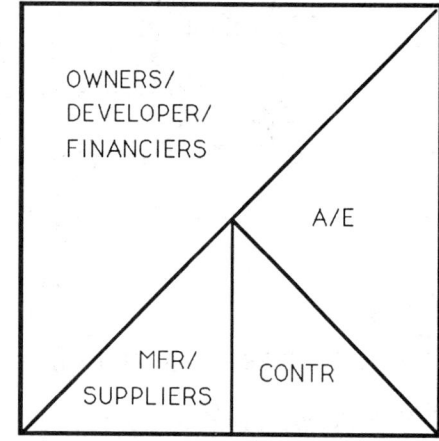

Figure 7: "PROFITABILITY"

Generally, the Owner/Developer ('D/O') and Financiers ('FIN') are not overly concerned with these 'Design-Build' arrangements, because they are still entitled to at least 50% of the total compensation for accepting only 3% of the liability responsibility, which is not a bad deal. Also, the Contractors ('CONTR'), Subcontractors ('SUBS') and Suppliers ('SUPPL') have more balanced and acceptable deals. Of major concern, of course (especially to many professional liability insurance carriers), is the overwhelming liability responsibility assumed by the A/E.

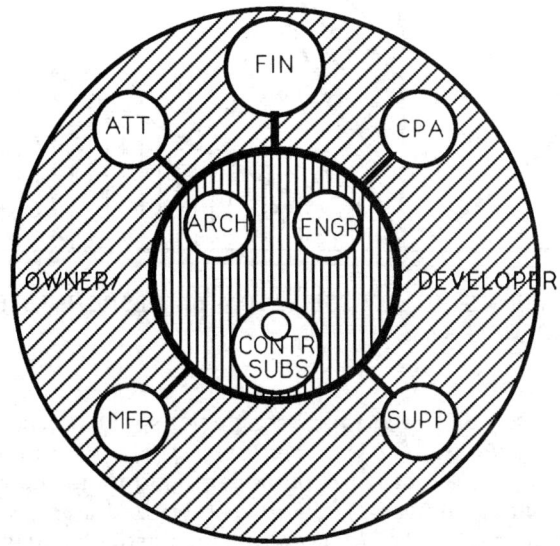

Figure 8: "DESIGN-BUILD-TURNKEY BUILDING CONSTRUCTION TEAM"

Figure 8 illustrates a recent development which has involved some very well-renowned A/E's and Developers on an international basis. For example, a major Hawaii Developer recently purchased an A/E firm that is providing him with 'in-house' professional design services. Some Owner/Developers have also purchased firms engaged as Contractors, and now they are in direct control of both material and product purchases directly from the Manufacturer ('MFR'). In these 'design-build-turnkey' building construction team arrangements, the Owner/Developers invariably are assisted by their own 'in-house' Certified Public Accountants ('CPA') and Attorneys ('ATT') who provide them with financial and legal counsel to keep them out of 'troubled waters'. And of course, they are intentionally organized to provide added real estate 'value' to their developments through improved 'quality control' assurance programs as I have noted in References 3 and 4.

Professional liability insurance carriers have devised an instrument of protection known as 'project insurance' to cover these progressive 'design-build-turnkey' contractural arrangements. Normally the term of coverage is limited to project completion and does not extend into the warranty period.

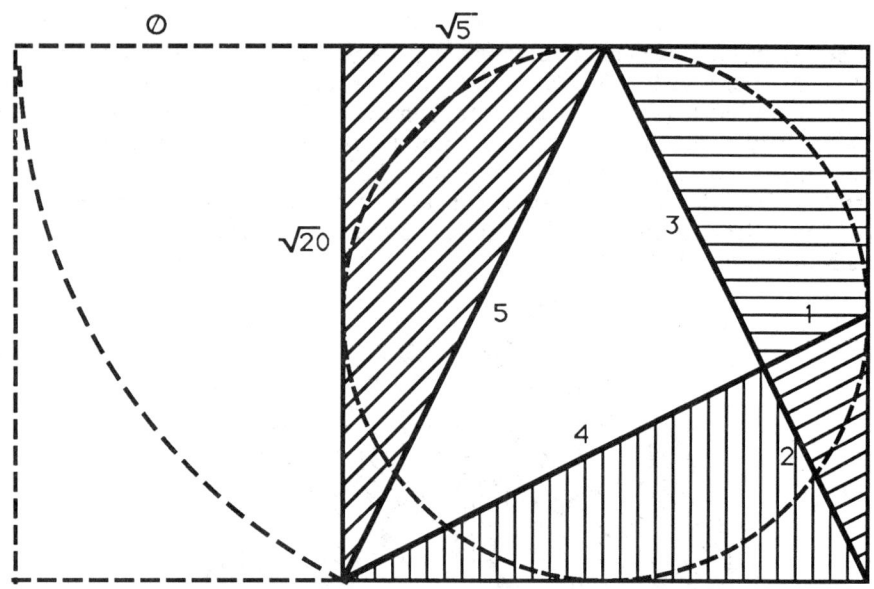

Figure 9: "SYNERGISTIC INTEGRATION of RESPONSIBILITY and PROFITABILITY"

Figure 9 is based on the author's conclusions which originated from personal 'architectural design research' efforts. It simultaneously represents a marriage between the 'Quadrature' of ancient Egypt and the reknowned 'Golden Section' of classical Greek geometries (civilized man's very first math and science, and the 'soul' of architecture). This particular diagram indicates how these harmonic proportions, which are a part of all living matter, can perhaps be a guiding light to synergistically integrate and harmoniously balance the levels of responsibility and profitability in any building construction project; and thereby, provide inherent 'quality control' assurance with the elimination of many construction deficiencies including water infiltration problems, and ensuing litigation. In Reference 5, I also mentioned the important role and contribution to the construction industry and design professions, which are afforded by the ASTM consensus standards process.

The area defined by unit lengths '1' and '2' (precisely defined by nature) belongs to the Owner/Developer, while the '2 x 4' goes to the material and product Manufacturers and Suppliers, and the '1 x 3' is the realm of the Contractor and his Subcontractors. The area defined by the '3/4/5' triangle of Pythagoras is naturally assigned to the Architects; and, of course, his Engineering consultants have the remaining area defined by unit length '5' (the radius of the 'Golden Section') and the 'square roots of 5 and 20'.

How to achieve an organizational structure and a viable economic communications network appropriate for the international urban marketplace, in order to aspire towards the matrical framework depicted in Figure 9, is a real challenge for the near future. Perhaps, Attorneys and Accountants would be willing to work with Owner/Developers, A/E's, Contractors, Manufacturers and Suppliers in order to achieve such noble purposes within the construction industry. Different contractual and fee schedule arrangements other than those depicted in this particular treatise may also be required. With professional compensation more appropriately matching risk-taking, greater attention would undoubtedly be paid to design details and quality assurance programs during construction as I noted in Reference 1.

In response to the perception of local needs in Hawaii eminating from a preponderance of water infiltration problems, property damages, and ensuing construction litigation and the need to collect and disseminate the research data and lessons being learned by 'privileged' members of the construction industry, I formed the 'Hawaii/Pacific CIRIES' organization within the Honolulu Chapter/CSI (Construction Specifications Institute) over three years ago as reported in Reference 6. 'CIRIES' is an acronym for 'Construction Industry Research, Information, Education Services'. CIRIES conducts interviews and seminars, and also publishes a tabloid insert 'CONTEXT 16', which is distributed to the Honolulu Chapter/CSI members via their monthly newsletter 'DIVISION 17'.

National case studies indicate that A/E's and their professional liability insurance carriers assume the bulk of the responsibility for design and construction deficiencies and remedial repair costs. CIRIES in conjunction with 'AEPIC' (Architectural and Engineering Performance Information Center) has provided insights on *performance standards* and *design criteria* for various sectors of the construction industry including 'design professionals'. CIRIES has demonstrated that in many instances the applications of building codes and design standards are absolutely minimal at best. In its seminars and publications, CIRIES has concentrated on 'generic' research data, without direct reference to specific products, product trade names, manufacturers, suppliers, or participants in various construction litigation cases.

Because of certain preponderant architectural and structural engineering design philosophies, as well as various innovative 'cost-savings' construction techniques, we have witnessed an onslaught of *water infiltration* problems with the design and construction of exterior building walls within the last two decades in Hawaii. In Honolulu, most of the highrise residential condominiums, commercial office towers, and resort hotel developments use reinforced concrete as a construction material in lieu of structural steel. In several instances, precast prestressed concrete members have been used in conjunction with poured-in-place 'in situ' concrete construction. The reinforcing steel connections and post-tensioning technologies employed can be very challenging indeed; especially to labor forces with limited skills.

Often, the reinforced concrete frame *design details* produced by professional A/E's are not in keeping with the slick and flashy colorful sales brochures prepared by the Developer, and residential condominium Owners often become disenchanted with unfulfilled expectations. When promises by the Developer do not materialize, and dissatisfied Owners are further disappointed by premature failure of building materials and manufactured products caused by 'water infiltration' through the exterior building walls or the roof, or flooding of basement parking garages, you can rest assured that ensuing construction litigation will follow, and that all of the parties involved in the construction as previously diagrammed, will be named in the legal action.

Many reinforced concrete highrise 'design details' produced by A/E's are actually more appropriate for industrial use, rather than commercial or residential use. This is because, the 'design details' are oriented around the structural concrete frame 'off-form (in situ) finish', which can be rather crude and irregular and non-uniform, attempting to serve as 'rough window openings', etc. Even in situations where the Contractor builds within the 'allowable tolerance limits' established by ACI(American Concrete Institute) codes and standards, these criteria may not be sufficient in themselves to prevent unsquare and out-of-plumb rough openings with varying diagonal dimensions. Furthermore, there is the problem of coordination of tolerances permitted on the approved 'shop drawings' by the Contractor, Manufacturer, and the A/E. For example, caulking gun sealant application specifications by the Architect, and instructions and warranties issued by the Manufacturer for proper use of their materials, and guarantees furnished by the Contractor and his Subcontractors may vary widely in 'allowable tolerance limits'.

In standard residential wood 'stud frame' construction, or even in the Modern 'post and lintel' wood construction, the 'design details' furnished by the Architect do not require the Contractor to rely on finishes produced by rough carpentry; but rather, the Contractor can depend on finish carpentry to trim the window openings. Consequently, in reinforced concrete frame highrise construction in Honolulu, the Contractor invariably has relied much too heavily on finishing the surface of the poured in place concrete with an abundance of gypsum plaster. The reliance on such unspecified means and methods of construction can certainly prove to be disastrous, especially with regard to window openings, installation of window frames and sealants, and 'water infiltration' problems in exterior building walls.

Where the Owner/Developers or their Construction Manager consultants, serving as so-called 'value engineering' experts, have not eliminated 'compensating channels' or 'subframes' in 'window-wall' construction via 'cost-savings', the 'allowable tolerance limits' can be better maintained. This is no guarantee however, that 'water infiltration' problems will be eliminated. This is because the Contractor may increase his tolerances and produce 'in situ' concrete of lesser quality in shorter time and for less cost, in consideration of being able to use 'subframes' in the 'window-wall'. The Architect and Structural Engineer consultant, as well as the Contractor, must coordinate 'quality control' measures during construction of the reinforced concrete frame to accommodate upward 'camber' in exposed prestressed and precast concrete members, downward 'deflection' in poured-in-place parts of the exterior building walls, and the phenomena of shrinkage or creep in the concrete columns. Structural frame movement due to dynamic wind and earthquake forces must also be considered in all of the connections and the attachments to the <u>exterior building walls,</u> including 'window walls', 'curtain *walls*', and *'veneer claddings'* of precast concrete, masonry, stone, metal, and glass, or plastics. The selection and specifications for *'building sealants'* are also of paramount importance in the successful working of <u>exterior building walls</u> against *'dynamic structural forces'* and *'water infiltration'*. This is especially true when A/E's are combining both precast and prestressed concrete members with poured-in-place reinforced concrete, which behave very differently as interdependant 'design elements' in terms of structural frame movement and continuity of force dissipation at the predetermined design articulation joints.

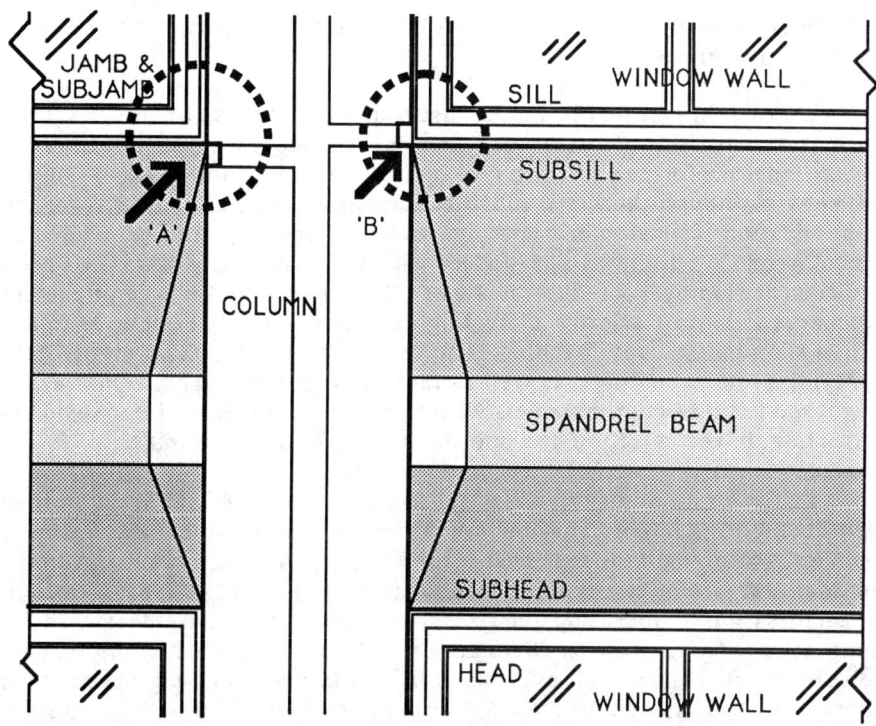

Figure 10: "RESIDENTIAL CONDOMINIUM 'WINDOW-WALL' FACADE"

Figure 10 illustrates a major 'water infiltration' problem due to the Construction Manager making revisions to the A/E's design solely with the Contractor, without consulting or notifying the A/E about the modifications in the field during construction. The circle and arrow showing detail 'A' on the left in accordance with the approved and accepted A/E design was <u>not</u> built; instead, the Contractor because of fabrication and erection problems elected to build detail 'B'. Depending on the spans involved, the precast prestressed concrete spandrel beams weighed between 4 and 7 tons each; and furthermore, *'rotational torsion'* was involved in these eccentric structural loadings and in the articulated 'mortise and tenon' connections. Consequently, for purposes of construction economy and erection safety, the Contractor modified the customized 'design detail' without due regard for 'water infiltration'. Moreover, the Contractor ignored the project specifications, and used gypsum plaster to attempt to plug the holes created by improper formwork in molding the recessed reinforced concrete column 'reveal'. Major damages accrued to the interior finishes including flooring and furniture of the occupants. Normally, such design revisions would be coordinated between the Architect and his Structural Engineering consultant along with the Contractor and his Glazing Subcontractor, and approved as a part of the 'shop drawing' review process. Then these 'shop drawings' would be followed and their implementation would be further coordinated in the field with all of the construction trades by the Contractor and the Construction Manager.

As shown in Figure 10, the precast concrete spandrel beam was sloped above and below the concrete floor slab line; consequently, without the use of traditional preformed 'drip strips', major weather staining occurred prematurely. Although there are specifications provided by Manufacturers for the application of elastomeric coatings on both horizontal and vertical surfaces of <u>exterior building walls,</u> different waterproofing specifications are required for sloping 'roof-like' surfaces. Elastomeric coatings can be 'tacky' and charged electrostatically, attracting environmental dirt particles in a fashion similar to a copy machine. The chemistry and biochemistry of material finishes and coatings, as well as factors of thermal expansion must be considered for different materials with varying coefficients of expansion and contraction, in the selection and specification of 'building sealants'.

As I have noted in reviewing and commenting on ASTM Main and Subcommittee Voting Ballots, there is much room for improvement in both ANSI (American National Standards Institute) and AAMA (Architectural Aluminum Manufacturers Association) Standards for the selection, specification, and certification of operating glass sash windows for highrise design and construction. 'Compensating channels' or 'subframes' (subsills, subheads, and subjambs) are rarely, if ever, used in certified window testing programs endorsed and sponsored by AAMA. Also, in lieu of structural steel or reinforced concrete frames, or masonry openings, the AAMA testing programs permit the use of wood frames, which swell with simulated sprayed rainwater in a manner similar to wood shakes on a roof. It is interesting to note, that when wood window frames were used in highrise construction prior to the use of metal frames, there were fewer 'water infiltration' problems due to the swell and natural sealant factor. Consequently, the AAMA test results are often invalid for the windows as installed, even though the windows bear the AAMA certification label.

In Reference 2, I reported on the results from case studies of 'windowwalls' and the major disparities between 'laboratory tests' of model window units and 'actual performance' of window and door installations in building construction. In Reference 7, I advised A/E's involved in the selection, specification, 'shop drawing' review and approval, construction observation and *'punch lists'* for installed windows and sliding glass doors, to have a thorough knowledge of the historical development in the design and manufacture of the windows and doors they are approving, just as it is necessary for an A/E *'design professional'* to have a thorough understanding of the historical development of prevailing code sections and design standards provisions.

Technically, the basic difference between *'window-walls'* and *'curtain walls'* is that the former fits between concrete floor slabs or between the top and bottom of spandrel beams as illustrated in Figure 10; whereas, the latter glazing system bypasses the floor slabs and spandrel beams as a distinct part of the exterior building envelope that is suspended from the structural system. This suspended anchorage(structural or non-structural) glazing system must take into consideration dynamic earthquake and wind force loadings, and allow for four-way movement systems, as well as code required fire resistive sealants. *'Curtain walls'* are much more sensitive to *'pressure differential'* between exterior and interior surfaces, which is most often exacerbated by air conditioning and mechanical ventilation systems, where rainwater is very naturally drawn into the building interior in some situations. Therefore, internal guttering systems are required in conjunction with the 'rain-screen' principle. The use of *'curtain wall'* glazing systems with precast and prestressed concrete construction can prove to be disastrous due to excessive movement in the structural frame and inability to control differential thermal movements and concomitant *'water infiltration'*.

In the 'window-wall' presented in Figure 10, three different unsuccessful attempts towards <u>permanent</u> *'remedial repairs'* on this and two other projects were pursued by purported industry experts in highrise window design and installation. In the process, it was learned that the Contractor had used gypsum plaster patch material extensively to treat *'honeycomb'* and *'rock pockets'* in the poured-in-place concrete column, as well as in the connection between the vertical columns and the horizontal precast concrete spandrel beams. Bronze anodized aluminum clip angles added to the exterior sub-frames to control the sealant bead width did <u>not</u> solve the *'water infiltration'* problem. Furthermore, drilling intermittent holes in the subsill track, and attempting to pack-fill the space with sealants recommended by the Manufacturer adjacent to the concrete rough window opening also failed due to air bubbles and inability to adequately clean the surfaces and space. Modifications to the *'weepholes'* in the sliding window sills only served to increase the *'pressure differential'* between interior and exterior environments and actually proved to be disastrous,in terms of accelerated damages to interior finishes and furniture, especially when the problem was exacerbated by the combination of high winds and rainstorms. Also, several ill-fated attempts to increase the interior sill heights to 'pond' excessive rain water penetrating the *'window-wall'* system ended in failure, and ensuing construction litigation was initiated by the apartment unit Owners.

In creating corporate *'signature'* and speculative highrise residential and commercial towers, Architects and Engineers have been very much involved in creating soaring *'roofless'* works of *'pinnacle'* architecture. These creative feats have exerted further pressures on sloping exterior building walls at the tops of highrise buildings where the 'wind speeds' and 'hydrostatic pressures' to withstand aggravated roof-like 'water infiltration' problems are the greatest. Most typical 'window-wall' and 'curtain wall' glazing systems are not designed or manufactured to be used in sloping conditions, unless they are customized. In many situations, 'design details' comparable to those used for 'skylights' are required; and very often, operating sash can no longer be incorporated within the sloping glazing system and fixed glass is the only sensible option. Standards and specifications for 'building seals and sealants' used in these particular applications are also of paramount importance, as are the review, coordination and approvals granted for the 'shop drawings'.

In earlier attempts to create 'roofless' highrise towers as with the outstanding Transamerica 'pyramid' in San Francisco designed by William L. Pereira, FAIA, Architect and other A/E's as shown in Figure 11, many mistakes were unwittingly made regarding expensively resolved 'water infiltration' problems. As vividly shown in Figure 11 below, special care in 'design details' is required to satisfactorily resolve the special challenges presented by sloping glazing systems, especially at the 'head' and 'sill' conditions of the 'window-wall' construction.

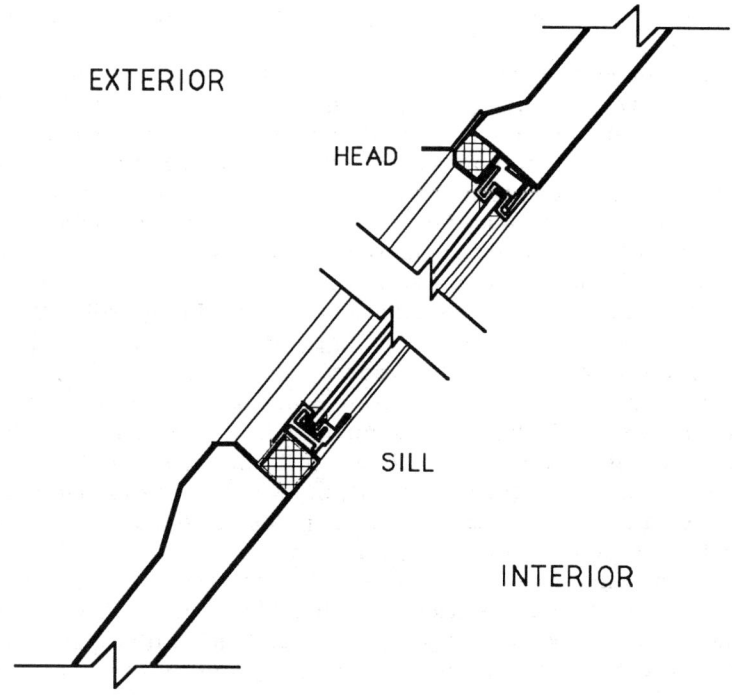

Figure 11: "PENTHOUSE APARTMENT 'SKYLIGHT' WINDOWS"

The crosshatched area shown in Figure 11, indicates the poured-in-place reinforced concrete 'design details' that the Contractor, in conjunction with the Construction Manager, decided to eliminate without consulting the Architect or the Owner. Their rationale, was that the Glazing Subcontractor wanted to install the 'skylight' portion of the 'window-wall' from the penthouse apartment interior, where the installation was not subjected to wind gusts. A metal 'rain diverter' flashing and additional 'sealants' were also installed by the Contractor when 'water infiltration' was experienced near the completion of construction and prior to occupancy. After responding to initial Owner complaints of severe 'water infiltration' during heavy rainstorms within the one-year warranty period, the Contractor opted to redo the concrete construction and the 'skylight' window installation in accordance with the original 'design details' provided by the Architect in consultation with the Glazing Subcontractor. For a number of obvious reasons, the installation did <u>not</u> work, and severe 'water infiltration' continued to plague the Owners with unsightly damages to interior finish materials and built-in cabinets and bookcases. The Owners became disappointed and disenchanted with the available experts, and retained their own experts and attorneys and filed a 'law suit'.

In this particular instance, 'shop drawings' were not issued by the Contractor or Glazing Subcontractor because the sloping penthouse apartment 'skylight' fixed windows shown in Figure 11, were not the product of a Window Manufacturer who would normally produce the 'shop drawings'. Instead, the Glazing Subcontractor opted to custom fabricate these window frames on the basis of a so-called 'stick system' used for more sheltered commercial storefronts. Therefore, the typical traditional design, fabrication and installation details were <u>not</u> coordinated with either the Contractor or the Architect and his Structural Engineering Consultant. Furthermore, after the concrete was poured and the forms stripped, the Developer decided that he wanted to improve on the design by the Architect, and admit more daylight to the dining rooms of the penthouse apartment units. Consequently, some concrete demolition work took place to provide wider rough window openings. Since the window frames had already been fabricated for the narrower openings, the Glazing Subcontractor simply adjoined an additional smaller window unit, and used a sealant to affix a bonze anodized coverplate without a gasket or continuous weld at the junction of the two jambs. This resulted in massive water infiltration problems in conjunction with the design and construction details shown in Figure 11.

In Reference <u>7</u>, I noted that he AAMA/ANSI Voluntary Specifications do <u>not</u> take into consideration, the simultaneous effects of wind and rain in combination, or wind gust factors; therefore, the 'design pressure' and 'water test pressure' and required sill heights will vary dramatically depending on regional 'basic wind speeds' specified by the local building code and the height of the highrise building tower. It is definitely prudent for the Owner/Developer and the A/E to conduct 'wind tunnel' studies for the <u>exterior building wall</u> of a highrise tower within its simulated urban context, prior to the selection, specification and installation of either a 'window-wall' or 'curtainwall' glazing system.

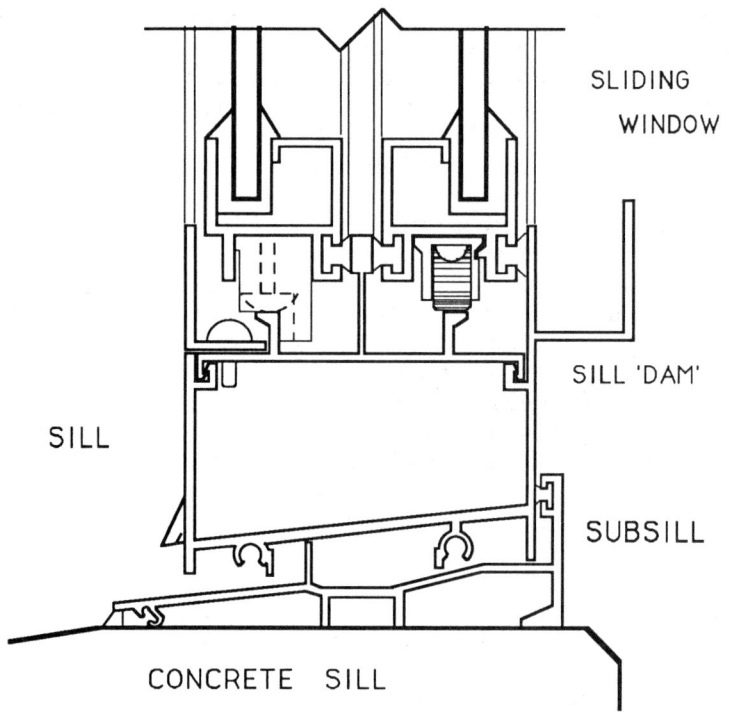

FIGURE 12: "HIGH PERFORMANCE SLIDING WINDOW SILL/SUBSILL"

The height of the sill 'dam' for sliding windows as illustrated above in Figure 12, is dependant on the height of the window above the street level. ANSI, AAMA, and the model building codes have adopted a 'design standard' of 30 feet above grade for the determination of 'minimum requirements' for 'basic wind speed'. The 'wind speeds', 'design pressures', 'structural test pressures', 'water test pressures', and other 'design factors' all vary with the height of the window above grade. Therefore, as they all increase with elevation above grade, the interior sill 'dam' ordinarily must also increase in height in order to prevent 'water infiltration' into the building spaces. As I noted in Reference 7, the 'High Performance' window and sliding door 'Load Tables' that are contained in the AAMA/ANSI Voluntary Specifications provide guideline criteria; however, most glazing system 'design consultant's' specifications exceed these 'minimum requirements'.

Since the interior sill 'dam' illustrated in Figure 12, functions as an internal rainwater gutter, it must be sealed at both ends; however, this particular 'design detail' is often overlooked in the installation of sliding windows on highrise construction. Even if not specified by the A/E, many Contractors and Glazing Subcontractors will install 'subheads' to compensate for downward 'deflection' and structural 'creep'; in reinforced concrete construction.

However, when 'subjambs' are also installed without interdependant 'subsills', the 'water infiltration' results can be disastrous. In these situations, the 'remedial repairs' may consist of either reconstructing the 'rough window openings', or modification of the window dimensions, or both. In several residential condominiums in Honolulu requiring extensive 'remedial repairs', the worth of the architecture and its immediate environment were devalued by insensitive 'remedial repairs' where selection and specification of inappropriate materials, colors and textures destroyed the original design expression by creative award-winning architects.

In Reference 3, I wrote that most Contractors are not insured for the quality of their workmanship; only, for 'consequential damages' (e.g., to carpets, draperies and furniture) caused by 'water infiltration', cracks in structural concrete or masonry construction. Also, unfortunately for most Contractors and other Defendant Parties involved in 'construction litigation" as outlined earlier, by the time a 'lawsuit' is filed, many of the original Subcontractors are out of business. In addition, there are Developers who limit their liability by forming 'shell' corporations that are project specific. Consequently, the A/E's and their professional liability insurance carriers become the 'deep pockets' as previously diagramed. In the long drawn out complex 'construction litigation' process, the professional liability insurance carriers can also become very frustrated with the mounting costs of litigation and lack of acceptable 'remedial repair solutions' and settlement offers. They may find themselves encouraging their insured A/E's to propose expedient and temporary 'quick fixes' in lieu of more permanent 'remedial repairs'. Unfortunately, this approach on highrises in Honolulu, and even on midrises and lowrise construction in other parts of the Hawaiian Islands, has generated the second and third wave of 'construction litigation' on the same project, and often by a different Owner.

'Remedial repairs' to existing structures that are occupied, is a highly specialized construction area, that needs to be addressed more poignantly in both our 'building codes' and 'design standards' in the very near future. This activity, whenever possible, should involve the original 'design architect' who best understands the rendered proportions and expressed emphasis on vertical, horizontal, diagonal, and curvilinear design elements, as well as the subdued elements, and the creation of mass and void in the exterior design elevations and 'design details'. However, in many instances, it must be realized that the original 'design architect' may be an artistically inclined 'primadonna', who is less inclined to want to accept less sensitive technological expertise. Therefore, it is often very helpful to involve the expertise of very good Architectural Engineers as I noted in Reference 3. Ordinarily, these Architectural Engineers are better able to bridge the gaps between artistic architectural expressions and functional technological needs involving the combination of construction techniques and the interface of various building materials and chemical compositions, as well as the proper application of 'building seals and sealants'. In this particular manner, the first, second and third waves of very expensive 'construction litigation' on the same project may be avoided. In addition, because of their engineering backgrounds, they are generally more sensitive to the use of toxic and hazardous building materials and volatile chemical solvents that are harmful to human beings, animals, plants, furniture finishes, and other aspects of our environments.

In References 3 and 4, I wrote on *'quality control'* in architectural design and building construction. It is interesting to note that not only the 'ARCHITECTURE' Journal published by The American Institute of Architects, but also 'PROGRESSIVE ARCHITECTURE', 'ARCHITECTURAL RECORD', and other architectural, building design, and construction periodicals and journals have placed an added emphasis on *'technology'*. This emphasis on *'technology'* is not limited to building construction *'quality control'* and Construction Management efforts reinitiated in the early '80's, but also includes essential aspects of basic architectural design considerations such as *'design detailing'* and *'specifications'* writing during the production of *'construction documents'*. It also includes a resurgent interest in more stringent review procedures including better coordination and review of *'shop drawings'* throughout the entire *'construction industry'*. In addition, as I noted in References 3 and 4, greater importance is being placed on *'construction administration'* and field supervision, observation, and code mandated *'special inspections'* of various structural items, and better scrutinized evaluations of building material and construction system *'test reports'*.

Due to the crisis caused by dramatically increasing professional liability insurance rates in the '80's, many large U.S. A/E firms have reorganized their firms to address the added responsibilities of Construction Management and *'quality control'* in building construction and in the preparation of *'construction documents'* as well as in conducting building code mandated *'special inspections'*. As a consequence, over 50% of the billings now for large A/E firms are not for professional design related services, but rather for Construction Management. Furthermore, as the volume and quantifiable costs of construction have increased, A/E's fees for professional design services have been trimmed on a percentage of construction cost basis, and have been further diminished by 'high technology' electrical and mechanical systems. In addition, *'fee schedules'* for A/E's have been outlawed by the U.S. Department of Justice, and A/E's have had to supplement their basic professional design fees with the comprehensive provision of Interior Design and Construction Management services. In the final analysis, Owners, Developers, and Consumers are paying more for added *'quality control'* measures now being implemented pervasively throughout the construction industry in Hawaii and elsewhere on a national and international basis. Consequently, as has been observed in ASTM Meetings, Symposiums, and in the 'STANDARDS' Journal, ASTM Standards Development in the area of Building Construction is moving beyond national *'design standards'* towards global and international *'design standards'*.

In Reference 1 and earlier in this treatise, I diagrammatically outlined the challenging needs to strongly consider reorganizing the *'construction industry'* including the more traditional and modified 'Triangular Contractural Relationships' presented in Figures 3 and 4, which have become fairly extinct for large scale urban design projects involving 'superblocks'. In order to further improve *'quality control'* results, as I have presented in Figure 9 on 'Synergistic Integration of Responsibility and Profitability' and in References 1 and 8, I believe it will be necessary in the very near future to harmonically balance liability responsibility and 'culpability' with 'profitability' on both large and small scale projects in the construction industry.

Because of their predominant involvement in creatively managing complex problem-solving, city planning, and large scale urban design projects, it is my belief that innovative Architects with the assistance of ingenious Engineers as noted in References 9, 10, 11, 12, and 13 must further posture themselves within society and the construction industry, to accept the paramount challenges which lie ahead. In order to accomplish this formidable goal, the A/E's, scientists, and technologists who are dedicated and committed to the improvement and development of our ASTM and other *design standards*, should exercise their very best professional judgement to uphold and maintain these Standards above the level of 'satisfactory' or *'minimum requirements'* contained within the 'construction industry' dominated Building Codes. In this manner, in conjunction with improvements in education and also the training of Architects, Engineers, and Contractors as I noted in Reference 3, the preponderance of 'water infiltration' problems and 'ensuing construction litigation' problems and costs can be eliminated, in order to make the U.S. Construction Industry more competitive in both the national and international marketplace.

REFERENCES:

[1] Yanoviak, A. C., "The Development Process and Building Failures", HAWAII ARCHITECT, Journal of the Hawaii Society / The American Institute of Architects, Vol.16, No.1, Honolulu, January 1987, pps. 18 - 25.

[2] Yanoviak, A. C., "Windows and Doors: Professional Practice Alerts", HAWAII ARCHITECT, Journal of the Hawaii Council / The American Institute of Architects, Vol.19, No. 12, Honolulu, December 1990, pps. 34 - 36.

[3] Yanoviak, A. C., "Quality Control in Building and Architecture"; Part I, HAWAII ARCHITECT, Journal of the Hawaii Society / The American Institute of Architects, Vol.17, No. 5, Honolulu, May 1988, pps. 36 - 39.

[4] Yanoviak, A. C., "Quality Control in Building and Architecture"; Part II, HAWAII ARCHITECT, Journal of the Hawaii Society / The American Institute of Architects, Vol.17, No. 6, Honolulu, June 1988, pps. 34 - 35.

[5] Yanoviak, A. C., "Water Infiltration Codes and Standards"; HAWAII ARCHITECT, Journal of the Hawaii Council / The American Institute of Architects, Vol.19, No. 5, Honolulu, May 1990, pps. 26 - 27.

[6] Yanoviak, A. C., "Hawaii/Pacific CIRIES"; HAWAII ARCHITECT, Journal of the Hawaii Society / The American Institute of Architects, Vol.16, No. 3, Honolulu, March 1987, pps. 38 - 40.

[7] Yanoviak, A. C., "Building Codes and Design Standards: Who's Responsible?"; HAWAII ARCHITECT, Journal of the Hawaii Council / The American Institute of Architects, Vol.19, No. 10, Honolulu, October 1990, pps. 25 - 29.

[8] Yanoviak, A. C., "Harmonic Proportions: Design Standards for Architecture"; HAWAII ARCHITECT, Journal of the Hawaii Council / The American Institute of Architects, Vol.19, No. 12, Honolulu, December 1990, pps. 28 - 33.

[9] Yanoviak, A. C.,"Architectural Design Challenges for Sealant Technology and Design Standards"; Building Sealants: Materials: Properties: and Performance, ASTM STP 1069, Thomas F. O'Connor, editor, American Society for Testing and Materials; Philadelphia, Pennsylvania 1990, pps. 334 - 346.

[10] Yanoviak, A. C., "Universe: City 2000; Geometrical Explorations of Urban Form and Content", Tall Buildings and the Growth of Cities: Pan Pacific Tall Buildings Conference Proceedings, Ernest N. L. Chiu and Ernest T. Yuasa, editors, Joint Committee on Tall Buildings:International Association for Bridge and Structural Engineering; American Society of Civil Engineers; The American Institute of Architects; American Institute of Planners; International Federation for Housing and Planning; International Union of Architects; Honolulu, Hawaii, January 1975, pps. 76 - 89.

[11] Yanoviak, A. C., "Master Builder or Master Planner"; HAWAII ARCHITECT, Journal of the Hawaii Chapter / The American Institute of Architects, Vol. 1, No. 9, Honolulu, Hawaii, September 1972, pps. 12 - 14.

[12] Yanoviak, A. C.,"City Planning and Urban Design as Architecture";Part I, HAWAII ARCHITECT, Journal of the Hawaii Chapter / The American Institute of Architects, Vol. 2, No. 7, Honolulu, July 1973, pps. 4 - 6.

[13] Yanoviak, A.C.,"City Planning and Urban Design as Architecture";Part II, HAWAII ARCHITECT, Journal of the Hawaii Chapter / The American Institute of Architects, Vol. 2, No. 8, Honolulu, August 1973, pps. 4 - 7.

David W. Kehrli

PREVENTION OF WATER PENETRATION THROUGH EXTERIOR DOOR SYSTEMS.

REFERENCE: Kehrli, D. W., "Prevention of Water Penetration Through Exterior Door Systems," <u>Water in Exterior Building Walls: Problems and Solutions, ASTM STP 1107</u>, Thomas A. Schwartz, Ed., American Society for Testing and Materials, Philadelphia, 1991.

ABSTRACT: Failure rates of exterior door systems are documented as they relate to water penetration in swing door systems. [1] Failure generally occurs because of improper accounting for pressure driven water in threshold and seal designs. Door bowing caused by wind pressure and thermal gradients can cause a severe loss of sealing efficiency by reducing the compression ratio on the weatherseal and by exposing cracks and joints in the frame/weatherseal interface. All of these elements can combine to cause door systems to fail in their responsibility in preventing water from penetrating into the interior of the building exterior envelope. This paper will detail the elements of proper door system design which is required for the prevention of water penetration. Issues to be presented are materials, components (seals, hardware and sills), design and testing.

KEYWORDS: design guidelines, door systems, hardware, weatherseal systems, test standards, threshold systems, high performance materials, components, and systems

<u>Exterior Door Watertightness Designs</u>

Exterior doors are designed to keep air, water, noise, dust, heat or cold out while keeping conditioned air in. They are also designed to keep the uninvited out as well as allowing the owner or users of the door systems to pass in safely with relative ease.

Mr. Kehrli is the Manager of Product Development and Fenestration Testing, Building Products Division, Schlegel Corporation - Rochester Division, 1555 Jefferson Road, Rochester, New York 14692.

As with many designs, some of these requirements conflict with one another or place overlapping demands on components and materials, often creating compromises on overall system designs. Designers tend to design within the mainstream of current practices and market stigmatisms. They look to performance specifications as "road signs" for proper designs as well as visiting the competition's line-up to provide assurance that they have designed a product which at least is as good or bad as the mainstream.

To examine this issue in greater detail, let's look at typical exterior door designs and how they present to the marketplace.

Door systems can be described as follows:

1. Wood, steel, aluminum, plastic, or composite frame. The frame comprises the head, jambs, mullions, and sill. The frame components can be "as is" or combinations of the above. Wood can be clad with plastic or aluminum to provide a maintenance-free exterior while leaving the interior material for end user finishing options; they can be left all natural, in the case of wood, for end user finishing. Aluminum can be painted or anodized. Plastics can be colored during the extrusion/pultrusion process. The options are endless.

2. Weathersealing systems can be complex, multidurometer, multimaterial, multiaction systems or simple; e.g. pvc bulb sealed. Exotic polymer seal designs are being designed and tested to achieve optimum, long-term performance.

3. Panels/slabs are insulated with various materials to achieve energy efficient designs. Glass infills are single glass or sealed insulating glass. They can be clear or low-E coated, inert gas filled, thermal or metallic spacer systems.

The panels are painted or clad. They are made from wood, plastic, aluminum, and steel. They can have internal stiffeners to reduce bow and deflection. They are made from built-up sections, laminates, and by-products of other materials; i.e., finger jointed, particle board, or Oriented Strand Board®.

4. Hardware systems are designed for operator/owner security and safety. They provide panel adjustability after installation. They allow operator friendliness, and low operating opening/closing force. They provide added stiffness for increased air and water performance.

Fenestration design theory suggests that window and door components; i.e., hardware, weatherseals, sealants, fasteners, glazing, framing members, etc., are assembled in such a manner that they successfully interact with and complement each other, thereby providing a quality product. Quality products, as defined here, are those products which meet all the requirements of the end-user. The proper selection of materials, designs, manufacturing techniques, and installation practices are all critical elements to the success of quality products.

Performance is defined by product designers, architects, product specifiers, builders, or anyone who has a need for products which must be designed to address issues of wind loads, water infiltration, installation, geographical considerations, ergonomic factors, and seasonal temperature profiles. Successful performance is a quality attribute that manufacturers strive for in their product designs and their approach to their markets. Performance is often misrepresented and misunderstood by the designer, specifier, and most importantly, the end-user of exterior door systems. Misrepresentation is due to "marketers eagerness" to sell minute differences in performance; i.e., 0.05 scfm/fcp vs. 0.07 scfm/fcp, and their lack of knowledge of materials, designs, theory, test procedures, in-service requirements and "actualities"; i.e., "real world performance".

Trade associations design and specify performance standards and test methods for these products. Typically, the standards are developed with off-the-shelf product performance capabilities in mind. This means that current materials, components and design concepts are used to define boundary parameters for which pass/fail or grade level ratings are established. It is rare that product performance specifications are developed for products which are unable to meet the standards. The industry typically ends up with performance standards which provide umbrella protection to the majority of the manufacturers. The marketplace ends up with products which often look alike and may be questionable as to their field performance and durability.

In order for fenestration products to meet the needs of exterior building envelopes for weathertightness, product designers and specifiers should address the following issues when developing quality exterior door systems.

ISSUES

Materials Of Manufacture

Improper selection of materials can affect stiffness, weathertightness, operability, design, security, durability of weatherseal functionality through weathering and user interaction, and use and abuse. Functionality is defined as the ability to maintain critical sealing properties; i.e., low compression set, low load deflection, low shrinkage, low wear and good weathering resistance, etc.

Sill materials can affect closing forces, compression ratios, adjustability, "wearability", sealing, color retention, and ergonomics.

Hardware materials can affect forced entry resistance, panel stiffness and bowing characteristics.

Weathersealing materials can affect air and water tightness, operational forces, chemical compatibility, color fastness, loss of strength and integrity, form, function and durability. [2]

Hardware Interaction With Weathertightness

Panels and slabs are affected by thermal gradients and wind loads which alter sealing dynamics of single vs. multipoint locking systems and lockset weathertightness.

Weathersealing Properties

Weathersealing systems are designed with the following properties which help door systems achieve overall envelope weathertightness: Low temperature flexibility, compression set, wear, weatherability, load deflection, shrinkage, coefficient of thermal expansion/contraction, water absorption, chemical compatibility, cleanability, and replaceability. Improper selection and design will cause door system failure as it relates to weathertightness. [2]

Frame/Panel Fastening Techniques

Framing techniques for the panel/slab to sills, header, and jambs are critical for they affect weathertightness, operability, and durability of system integrity and materials. Astragals and mullions are critical locations for weatherseal interactions. Mechanical fastening systems, set-up jigs, and placement of frame members to fastening points are critical to proper installation and long-term performance.

Installation Techniques Of Door Systems Into The Exterior Building Envelope

The installation is critical for weathertightness, operability, and proper fit. Improper installation techniques often negate the best of designs by altering critical dimensional relationships between structural, panel, and sealing systems. Unless adjustability factors are designed into the components and assemblies, field repairs and "fixes" can be expensive.

Fabrication Techniques

Maintaining reveals, mounting distances, setbacks, etc. are absolutely crucial for weathertightness. Weatherseal joinery, sealing of frame and cladding joints, fastening/joining devices, and hardware integrity are areas which cause door systems problems as they are transported, installed, and used throughout their design life. These areas, when subjected to stress, may result in water tightness failures.

Threshold System Integration

The integration of the threshold to the frame, slab, and weathersealing systems are critical to weathertightness, operability, and durability of performance and materials. Materials used in the manufacture of sill systems will affect weathertightness, fabrication methods, design integrity, coefficient of expansion, ergonomics, and esthetics.

Environmental And Geographical Considerations

Effects of wind pressures, temperature, humidity, rain, freeze/thaw, ozone, pollutants, uv, salt spray, building designs, builder techniques, building codes, and local materials all contribute to weathertightness, operability, and cost.

User Interactions

The owner/user of building fenestration systems impose traffic loads, maintenance schedules, proper finishing, use and abuse factors, and normal wear and tear.

Test And Performance Standards

Issues which must be addressed by designers are the appropriateness of the specified standards and test methods for the specific job site and products under test or evaluation. The designer needs to know the background of the test standard being specified for product compliance, what products, materials, and parameters are applicable to the standard, what deficiencies are present in the standard; i.e., what are the limitations and tolerances on the interpretations of the data obtained from the standard. How can the designer improve upon the test standard in order to refine the overall performance attributes of the product(s) in question? These issues greatly impact the final form and function of door systems.

Other tests are available to describe door system performance; u-value, CRF, differential temperature air and water leakage, accelerated aging, and sound transmission loss. Designing to these requirements will add performance, value and cost.

Maintenance Free Finish Designs - Claddings

Cladding deficiencies can affect weathertightness and materials durability. Areas such as sealing techniques and materials, location of joints, and types of joints can affect maintenance-free finishes.

Atypical Leakage Locations

Component joinery and tightness are crucial elements in watertightness of door systems. Typical areas of water leakage get attention to sealing while non-typical or misunderstood areas are neglected. It is these areas which can cause the greatest water damage after installation; i.e., lock systems, frame joints, and seal joints.

Operation Theory

Operation of door systems and their components depend on proper training and instruction sets to end-users and designers. Adjustable hinges, door sill saddles, door bottom seals, sweeps, and locksets can all contribute to improved weathertightness. These require manufacturer, builder, or operator/owner intervention to resolve field leakage problems. Often the adjustments are not enough to fix the problem permanently, or after adjustment a new problem arises. This is typically due to overadjustment and subsequent over stress and failure of other critical components.

Design And Theory

Many design principals are employed to keep water from entering the exterior building envelope and its components. Some of these are rainscreen, pressure equalized design, pressure head, filtration, run-off, water management, and "zero-leak" systems. These principals in conjunction with materials and designs provide "quality" door systems with long-term performance. When these principals are ignored or improper materials are chosen for the components, field failures are inevitable.

Durability, Materials And Performance

The durability of selected materials and designs can drastically affect the installed performance of door systems. Laboratory tests to measure fenestration system durability are being developed at ASTM. These tests will aid the designer or specifier in selecting materials and components which have a propensity for long-term performance. In the meantime, designers must evaluate materials and designs based on suppliers' test data, verification tests, component and systems tests, and statistical extrapolation.

Manufacturers often resist providing performance warranties and guarantees due to the installation/installer variables which can affect materials, components, and designs. Manufacturers are also unable to control where their products will be installed. This may have adverse affects on the selected materials due to environmental degradation factors, user stresses, and misuse.

Serviceability

Designing for long-term performance requires complete definition of the environmental degradation factors which can lead to material failures. It requires proper accelerated aging test methods which allow material and design evaluations, ranking of materials and designs, and evaluation of overall system integration performance.

Cost Factors

Each element described above has a cost factor associated with the level of performance required. Material and component selections are based on marketability of performance and designs. Costs are balanced against market performance demands. Costs, however, do not dictate total performance, as materials and design tradeoffs commonly lead to acceptable compromises for the required envelope design. Each of these issues must be tempered with the needs and wants of the marketplace, the end-user, the owner, and the building itself in order to achieve cost effective and quality designs.

Recommended Design Strategy

Door system technology has come a long way in providing quality products. The question which needs answering is, "Why do some systems fail and others do not when installed in the exterior envelope?". Is laboratory performance really obtainable in the field? How can door manufacturers improve their products without pricing themselves out of the marketplace? These answers may be found in the following design strategy.

1. Strictly adhere to basic, fundamental laws governing fluid flow. [3]

2. Keep designs simplistic and pure.

3. Specify performance-oriented materials and components which meet long term door system performance needs.

4. Fully understand the principles behind the tests used to simulate real-world performance in the laboratory. [3]

5. Test every design, material, and component for singular and interactive reactions to real-world environmental and user parameters.

Design Guidelines

Quality door products can be achieved in the manufacturing process and the marketplace by addressing these design guidelines.

1. Panel stiffness **must** be great enough to maintain proper weatherseal compression ratios during wind loading and thermal gradients. Decreasing weatherseal compression ratios will allow increased air, water, noise, dust, heat or cold infiltration/exfiltration.

2. <u>All joinery must be sealed!</u> <u>No exceptions!</u> Water and air penetrate cracks, joints, holes, etc. at laboratory test conditions. The effects are worsened during thermal gradient conditions, simulated or "real-world". What was sealed at

laboratory or manufacturing temperatures are questionable at elevated or depressed installed conditions throughout the seasons.

3. Primary weathersealing should occur in full-compression modes, when possible. It should be uniplane in location and continuous whenever possible. The seal should always be designed with mechanical leafing action, i.e., these types of seals have extended reach without causing excessive closing force, to provide continuous contact with mating surfaces during panel/frame separation due to wind and temperature action. Constant, or near-constant compression ratios are critical for weathertightness.

4. Design for absolute weathertightness under all possible conditions, or, design for water management. <u>Nothing else is acceptable</u>. Water management theory states that water penetration is probable. Understand the factors which affect penetration. Fully accept the fact that keeping water completely out of the product or installation may not be obtainable and managing water penetration follows good building design practice. Management allows for penetration, but rejects product failure. Failure is defined as water in places which can cause or has caused damage or deleterious side effects to all other building materials, components, and systems within the building structure. [3]

5. Design for installed typical worst case scenarios, not specification levels or grades. What passes in the laboratory for 15 minutes never seems to pass in the field, under fully exposed conditions. And since we cannot control the location or building design in which the product may end up, it is wise to anticipate a "naked" installation; i.e., no overhangs or storm doors.

6. For low-stiffness panels and/or high wind load applications, multipoint locking systems, full perimeter uniplane weatherseals, and self-draining water management thresholds and flashings should be designed into the door system to obtain high performance, quality designs.

CONCLUSION

To achieve "watertight" door designs requires commitments to "quality". The cost of neglecting quality designs are field failures, call-backs, lawsuits, and loss of market share.

The elements presented in this paper can lead the product designer/specifier to products and producers which utilize sound principals in water management techniques.

ACKNOWLEDGEMENTS

The author acknowledges with much appreciation the help of Terry Warner for relentless patience in manuscript preparation; and Jim Burrous for valued qualitative interpretation of real-world problems.

REFERENCES:

1. Test Reports, Schlegel Corporation, Rochester, New York 14623, 1987-1990.

2. Kehrli, D.W., "Weatherstrip Design Guidelines", paper presented at National Wood Window and Door Association summer meeting, sponsored by National Wood Window and Door Association, Des Plaines, Illinois, 1988.

3. AAMA Window Selection Guide, American Architectural Manufacturers Association, Des Plaines, Illinois, 1988

Axel R. Carlson

COMPUTER SIMULATION OF WALL CONDENSATION PROBLEMS

REFERENCE: Carlson, A. R., "Computer Simulation of Wall Condensation Problems," <u>Water in Exterior Building Walls: Problems and Solutions, ASTM STP 1107</u>, Thomas A. Schwartz, Ed., American Society for Testing and Materials, Philadelphia, 1991.

ABSTRACT: Field reports from colder climates such as Alaska, Minnesota, Ohio, etc. indicate that wall moisture is a major problem, if there is no vapor retarder. A polyethylene vapor retarder is recommended in Alaska. Studies of retrofitted walls in older residences from warmer climates such as Oregon intimates that moisture in the wall may not be a problem even though there is no vapor retarder. However, the Oregon study still recommends a polyethylene vapor retarder for new construction.

There are two potential vapor retarders, which are designated in this paper as the primary and secondary vapor retarders. The primary vapor retarder for heated buildings, in cold climates such as Alaska, should consist of a polyethylene membrane placed at the inner skin directly under the gypsum wallboard (drywall). If the primary vapor retarder does not function as intended, then the outer skin consisting of a combination of plywood sheathing, siding, wind barrier, etc., may function as a secondary vapor retarder, which may trap moisture in the wall cavity.

Computer spreadsheet simulations of various wall sections indicate that wall moisture problems are dependent on inside and outside psychrometric conditions. The vapor pressure differential of the inside and outside air determines the driving force of the vapor. The vapor pressure and permeability of the primary and secondary vapor retarders determines how much water vapor may diffuse into the wall cavity. If the temperature gradients at the vapor retarder remains below dew point the moisture will condense in the cavity. If the temperature in the cavity remains below 32°F for prolonged periods the condensate will freeze and continue to build up as frost.

KEYWORDS: Condensation, vapor retarder, ambient temperature, dew-point temperature, relative humidity, absolute humidity, humidity ratio, vapor pressure, temperature gradient, permeability.

Mr. Carlson is Professor Emeritus, Extension Engineer, Cooperative Extension Service, University of Alaska, Fairbanks, Alaska 99755.

Numerous books and technical reports have been written on methods of preventing condensation-related maintenance in buildings, yet the problem still persists. The simplest and most illustrative book on the design of design of insulated buildings to prevent condensation-related maintenance problems that I have reviewed was written by T. S. Rogers, architect, in 1951, [1]. The most comprehensive engineering manual on the design of insulated buildings was developed by ASHRAE, [2].

Agricultural production buildings usually operate at relative humidities of 60 to 90 percent; hence, are often subject to severe condensation-related maintenance problems. An excellent text book on moisture control for farm buildings was written by Barre and Sammet in 1950, [3]. This book was revised by Barre, Sammet and Nelson in 1987, [4]. The relative humidity of modern energy efficient human housing may exceed 60 percent, if adequate ventilation is not provided. Therefore, these texts are also applicable to the design of human housing and other commercial buildings.

Moisture problems in buildings appears to be an international problem, [5-6].

<u>Minnesota Field Study</u>. In 1986 Minnesota investigated moisture problems in walls occurring in a series of panelized homes built between 1970 to 1985, Figure 1, [7]. A computer simulation of dew-point conditions in these houses will be used later on to determine why the moisture occurred.

Nine of the houses were examined in more detail by State of Minnesota and University of Minnesota teams. The houses with most significant moisture problems were further surveyed with the assistance of county agents. Indoor relative humidities ranged from 46 to 61 percent, with an average of 52 percent. Condensation occurred on most wood storm windows. Mold growth occurred on interior wood sash. Water stains occurred on the interior face of wall and ceiling coverings. Condensation occurred on basement joists. Also, there was general moisture stains on exterior siding, etc.

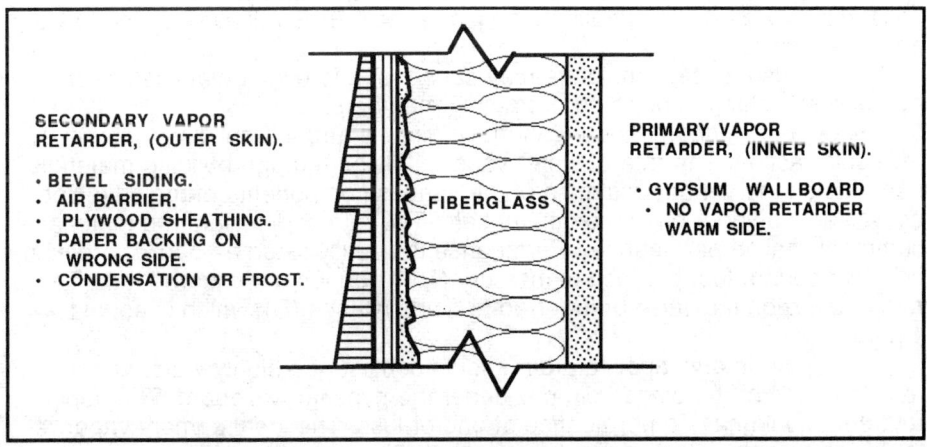

FIGURE 1. WALL SECTION (MINNESOTA HOUSE)

On April 17, 1987 I personally examined one of the Minnesota houses, with the principal investigator. The hardboard horizontal siding on the east gable of the cathedral roof was removed. The asphalt coated wind barrier between the sheathing and siding had disintegrated and appeared to be charred as from spontaneous combustion. The plywood sheathing was water soaked and so badly deteriorated it could be easily ripped out with ones fingers.

The asphalt coated paper vapor retarder backing of the fiberglass insulation of the Minnesota house had been placed on the cold side of the wall instead of the warm side. The tabs were stapled onto the wide faces of the studs rather than over the narrow faces. Even if the paper backing had been placed on the warm side, the air-vapor retarder would could not function as intended, because the tabs did not extend over the framing members. The tabs must be pressed into place on all framing members by the gypsum wallboard to be an effective vapor retarder. Further, all seams should be vapor sealed with tape.

Apparently moisture that diffused into the wall condensed out and continued to build up as frost. This has been observed in prolonged subzero climates such as Alaska. Moisture diffusing into the wall cavity will freeze at the secondary vapor retarder, if its temperature drops below 32 F, [8,9]. Similar occurrences have been observed in Maine, [10]. Moisture will continue to build up as ice on the inside face of the sheathing. When the ice melts during a spring thaw, a large of amount of water will be released suddenly.

The following wall problems have been observed in Alaska due to accumulation of excess moisture in a wall:

1. Ugly water or mildew stains may occur on the inside of wall or the carpeting, particularly if there is no vapor retarder.

2. Water may leak out the exterior of the wall and stain the siding.

3. Moisture trapped between the sheathing and siding, may refreeze causing wood, aluminum or plastic siding to bulge. It may even cause the wood siding to crack.

4. Over a prolonged period of time, excess moisture around siding nails may freeze causing nails to be jacked out. This will allow siding to warp and even crack.

5. Also, excess moisture may cause rot and even delamination of plywood sheathing and rotting of framing members.

<u>Vapor retarder</u>. Water can move directly through building material by molecular action. The rate of water vapor diffusion through building materials depends on the vapor resistance of the various components of the envelope, as well as the vapor pressure differential. The rate of diffusion thru material is commonly called permeance. Permeance is usually listed as perms, grains per hour per square foot per inch of mercury (70.7 pounds per square foot). The rate of diffusion may also be referred to as resistance (D), which is equal to 1/perm.

The primary vapor retarder should consist of a highly water vapor resistant membrane placed directly under the gypsum wallboard. The vapor retarder should have a permeance of one or less. The ideal primary vapor retarder commonly selected is a 6-mil polyethylene membrane. It has a water vapor permeance of 0.06 perms, (resistance 16.67).

Plain gypsum wallboard has a permeance of 50.0 perms (resistance 0.020), [2]. According to laboratory tests at North Dakota State University on various vapor retarders, gypsum wallboard vapor with a resistant paint only had a permeance of 7.35 perms (resistance 0.136), [11], while 6 mil

A single 1/4 inch diameter hole in the center of 6 mil polyethylene in a 16 square foot area resulted in a increase in permeance from 0.06 to 0.29 perms, (resistance 16.67 to 3.45), [10]. The effect of electrical outlet boxes on air-vapor leakage has been checked with blower door and smoke tests. Plastic pans are now available to facilitate the sealing electrical outlet boxes against air-vapor leaks.

The asphalt coated back up paper on blanket insulation has permeance of 0.40 perms (resistance 2.50) [2], if properly vapor sealed at all edges and seams.

<u>Wind barrier</u>. Many codes recommend that a wind barrier placed over the sheathing should have a permeability of 5 perms or greater (resistance of 0.20 or less). Asphalt coated sheathing paper used in many buildings has a permeability of 1.79 (resistance 0.56), [2], which is too low for a modern energy efficient building.

A wind barrier is not necessary in a modern building, if a polyethylene membrane is used as a primary vapor retarder.. The wind barrier has a tendency to retard the diffusion of moisture trapped in the wall. A wind barrier was required in older homes with shiplap sheathing to minimize heat loss due to excessive heat loss due air leaks through the walls.

Some experts recommend that a house be rapped with a micro-perforated 3-mil polyethylene wind barrier with a permeance of 0.176 perms (resistance 5.68). It should only be used with horizontal lapped wood or metal siding, where there is no sheathing to prevent wind driven rain from penetrating into the wall cavity.

The plywood sheathing often used on the exterior of the wall as bracing and back-up for horizontal siding could serve as the secondary vapor retarder or a wind barrier. It has a permeability of 0.35 perms (resistance 2.86), particularly when glued to the framing such as with a mobile home. Its permeability is less than the accepted standard of 5 perms for a wind barrier.

<u>Air changes</u>. The need for energy conservation has resulted in substantial reduction in air leakage and mechanical ventilation rates. In residences this was accomplished by various energy saving techniques such as weatherstripping, caulking, vapor retarders, wind retarders, etc. Natural air infiltration has been reduced to a low of 0.20 Cph. This air change has reduced heat losses in houses from a high of 24 to 38 percent to low of 11 percent. Also mechanical ventilation rates in public buildings has been reduced to conserve on energy.

One of the more important challenges in a modern energy efficient house is the control of excess moisture in the air. Excess moisture produced by respiration of humans, pets, plants, bathing, laundering, and manufacturing processes etc., must be removed by mechanical ventilation. Heat recovery ventilation (HRV) systems are now available to reduce air exchange heat losses.

Buildings built prior to the 1970 energy crisis usually had sufficient air changes (infiltration) which maintained the relative humidity and resultant vapor pressure at a lower level than modern energy efficient buildings. Further, the porosity of the exterior skin allowed excess water or ice tn the cavity to evaporate or sublimate fairly easily. The tightness of the plywood sheathing, siding, wind barrier has reduced air-vapor leakage.

Natural air exchanges of 0.50 to 1.50 Cph (changes per hour) have been measured in houses built prior to 1945. This resulted in an estimated 24 to 38 percent of the total heat loss.

A super-insulated home in Fairbanks had a blower door air leakage test of 5.48 Cph (changes per hour) at 50 Pascals. This may result in an average natural air change of 0.49 Cph. A 24-0x48-0 1-story house with a day-light basement may have a natural ventilation rate of 160 Cfm (cubic feet per minute). This meets ASHRAE recommendation of 0.50 Cph.

Mechanical ventilation is essential to maintain an air change of 0.50 Cph as recommended by ASHRAE (American Society of Heating, Refrigerating Air Conditioning Engineers). This air change is necessary to provide fresh air for breathing and to remove excess water vapor in the air. Also, it is needed for the dilution of disease producing organisms and carcinogens. Further, an adequate supply of air is essential for efficient combustion of a fossil fired heating system and to prevent the formation of carbon monoxide combustion gases. Therefore, a fossil fired heating plant should be provided with a separate air intake.

Air Drywall Approach. The air drywall approach (ADA) has been suggested as a technique for minimizing heat losses due to air leaks, [12]. The ADA technique consists of installing a gasket over the framing members before the drywall (gypsum wallboard) is installed. Some experts maintain it would reduce vapor leaks as well. However, field studies indicate that water vapor can still diffuse through the gypsum wallboard and other wall coverings, if it has no vapor retarder..

It would be easier to visually inspect the workmanship of the polyethylene vapor seal, electrical wiring, plumbing, etc, before the gypsum wallboard is installed. Both air and vapor leaks can be minimized at less labor and material costs by an adequately sealed polyethylene vapor retarder. ADA would require a blower door test to verify the adequacy of the air seal. Unfortunately, the air door test only verifies air leaks and does not test the permeability of either the gypsum wallboard or the vapor retarder. The durability of a painted vapor retarder in an earthquake zone is questioned as gypsum wall board may crack if the building shifts due to an earthquake or settlement due to permafrost.

Wall vapor retention. A 1963 University of Maine research report on a series of insulated test panels and with various vapor retarders during the winter of 1961 indicated that a certain amount of moisture and frost formed on the inner face of sheathing, [10]. The walls were insulated with mineral wool (fiberglass). Also the report indicated that a certain amount of moisture was retained in the wall cavity. The test wall panels were covered on the exterior with 3/8" exterior grade plywood. The interior was covered with 3/4" square edge pine boards, which should not deter the passage of vapor retarder to the primary vapor retarder.

A 6 mil continuous polyethylene vapor retarder placed under wood boards allowed 0.070 perms to enter the test panel, while 0.006 perms left, such that 9 percent was retained. A continuous 4 mil polyethylene vapor retarder allowed 0.130 perms to enter the test panel, while 0.096 perms left, such that 26 percent was retained. A 4 mil polyethylene with 5/16" staples 2" o.c. into studs and 2' o.c.tape over 2 inch lap joints allowed 0.128 perms to enter, while 0.108 perms to left, such that 16 Pct (percent) was retained.

A 4 mil continuous polyethylene vapor retarder with a single 1/4" round hole centered in a 32 square foot panel allowed 0.321 perms to enter the panel, while 0.275 perms left, such that 14 Pct was retained.

Pressure and smoke tests revealed that electrical outlets and other penetrations in a wall or ceiling and floor may result in substantial air-vapor leaks. As a result, special plastic boxes have neen developed that can be sealed to the vapor retarder to eliminate air-vapor leaks. Electrical outlet boxes are placed in the plastic box. Electrical wiring that penetrates the box are sealed with a suitable caulking compound.

COMPUTER SIMULATION

A computer simulation of temperature gradients and dew-point conditions of the Minnesota houses was developed to determine why moisture occurred in the wall, Tables 1-6. Similar studies have been developed to analyze moisture problems in floors and roofs. Also studies have been conducted for buildings that are cooled by refrigeration during the summer season.

It would be extremely tedious, if not impossible, to develop this analysis without the aid of the modern computer. Perhaps this is why the cause of condensation problems still seems to be such a mystery to so many? These particular spreadsheets were developed on a Macintosh Plus computer using a Microsoft Excel spreadsheet.

The original formulae are listed in regular algebraic format, ASHRAE [2], The formulae were revised to a single line computer format to accommodate the electronic computer, D.B. Brooker in 1966 [13]. The formulae have been modified further by this author to accommodate the greater memory of the modern computer, Table 1.

Special symbols were devised for Tables 6.1-6.3. For example the value $Tg.n$ is taken from the same line, $Tg.n-1$ is taken from previous line, and $Tg.n+1$ is taken from the next line. All the formulae in the spreadsheet are visible when viewing respective cells in the spreadsheet.

For those who are not familiar with these type of formulae, a copy of a disk may be obtained by writing to the author. Similar spreadsheets have been developed for the IBM compatible computer using Lotus 123 and the IBM Microsoft Works spreadsheets. A summer cooling simulation will be included. A set of instructions will be included on a "README" file. There will be a nominal fee of $35.00 to cover the cost of the disk, packaging and mailing. Information on file system and format is required, such as (Mac - HFS, MSDOS, PRODOS, 400K, 800K, 1440K, 2400 K, 3.5" or 5.25").

English units of measure were selected as it is more understandable to the general construction industry of United States. The formulae could be converted to SI units of measure, if desired. Conversion factors for English to Metric (SI) units are listed in Table 2.

Mean Temperatures and Number of Months Below 32°F, Table 3. Studies were cited that frost (ice) was found in the wall cavity. Therefore, it seemed appropriate to explore how many months the mean temperature could remain below freezing. The mean temperature and number of months it occurred at Anchorage AK was 5 months, Cleveland OH 3

TABLE 1. TYPICAL FORMULAE

Name	Units	Symbol	Formulae
1. Ambient temperature	°F	T	"=Given
2. Relative humidity	Pct	Rh	"=Given
3. Vapor pressure	Psf	Pv	"=(IF(C3>32,(EXP(54.6329-(12301.69/ (459.7+C3))-(5.16923*LN(459.7+C3)))*144, (EXP(23.3924-(11286.65/(459.7+C3))- (0.46057*LN(459.7+C3)))*144)*Rh/100
4. Absolute humidity		Hv	"=0.00431875*Pvi/(14.696-0.006944*Pvi)*7000*Rh/100
5. Dew-point temperature		Td	"=IF(T>32,79.047+30.579*LN(Pv*0.014139)+ 1.8893*(LN(Pv*0.014139))^2, 71.98+24.873*LN(FPv*0.014139)+0.8927* (LN(Pv*0.014139))^2)
6. Temperature Gradient	°F	Tg.n	Tg.n-1+((Ti-To)/Rt*R.n)
7. Pressure Gradient	Psf	Pd.n	"=IF(Tg.n>0,Pv.n+1-((Pvi-Pvo)/Dt*D.n),Pv.n+1)
8. Vapor Diffusion	Gr/Hr Sf	Vd.n	"=IIF(Tg.n>0,1/D.x/70.3*(Pv.n+1-Pv.n),Pv.n+1)
9. Dew-point Temperature	°F	Td.n	"=IF(Tg.n>32,79.047+30.579+LN(Pv.n*0.014139)+ 1.8893*(LN(Pv.n*0.014139))^2,71.98+24.873* LN(pv.n*0.01439)+0/829*(LN(Pv.n**0.14139))^2)

TABLE 2. CONVERSION FACTORS (ENGLISH TO SI UNITS).

Item	To covert from	To	Multiply by
Area	ft^2	m^2	9.290 304E-02
Area	in^2	m^2	6.451 600E-04
Energy	BTU	J	1.054 613E+03
Energy	kWh	J	3.600 000E+06
Length	ft	m	3.048 000E+01
Length	in	m	2.540 000E-02
Mass/volume	lb/ft^2	kg/m^3	1.601 846E+01
Mass/volume	lb/in^2	kd/m^3	2.767 990E+04
Power	Btu (IT)/hr	W	1.758 427E-01
Power	Btu (IT)/min	W	3.766 161E-04
Pressure	bar	Pa	1.000 000E+05
Pressure	atm. (standard)	Pa	1.013 2506E05
Pressure	in. Hg (60 F)	Pa	3.376 850E+03
Pressure	in.H2O (60 F)	Pa	2.488 400E+02
Pressure	lb/ft^2	Pa	4.788 026E=01
Pressure	in/ft^2	Pa	6.894 757E+03
Resistance, Thermal	R	RSI	
Resistance, Vapor	D	DSi	
Temperature	F	C	(F-32)*1.8
Temperature	C	K	(F+459.67)1.8
Temperature	R	K	
Volume	ft^3	m^3	3.523 907E-02
Volume	in^3	m^3	1.638 706E-05

TABLE 3. MEAN TEMPERATURES AND HEATING DEGREE DAYS OF SELECTED CITIES

Month	Jul	Aug	Sep	Oct	Nov	Dec	Jan	Feb	Mar	Apr	May	Jun	Annual Mean	Months 33°F & Colder	Degree Day 65 F
Month Number	7	8	9	10	11	12	1	2	3	4	5	6			
Days in Month	31	31	30	31	30	31	31	28	31	30	31	30	365		
Anchorage, AK	57	56	48	35	22	14	12	16	23	36	46	55	35.0	5	10,950
Billings, MT	65	65	59	49	35	28	23	26	34	46	56	62	45.6	3	7,081
Boise, ID	65	65	61	52	40	32	29	35	42	50	57	62	49.1	2	5,804
Buffalo, NY	64	64	60	51	39	28	24	24	31	44	54	62	45.5	4	7,118
Cleveland, OH	65	64	62	53	40	30	28	28	35	47	57	63	47.5	3	6,388
Dallas, TX	65	65	65	63	54	48	46	49	55	62	65	65	58.5	0	2,373
Denver, CO	65	65	61	51	38	32	28	32	36	46	56	63	47.7	3	6,283
Duluth, MN	63	61	54	45	27	14	9	11	21	37	49	58	37.5	5	10,038
Erie, PA	65	64	62	52	41	31	27	26	34	46	56	63	47.2	4	6,451
Fairbanks, AK	59	54	44	26	4	-8	-11	-3	9	29	63	58	27.1	7	13,834
Fargo, ND	64	64	58	46	28	14	7	11	24	42	54	62	39.6	5	9,271
Green Bay, WI	64	63	59	49	34	22	17	18	28	43	54	62	43.0	4	8,029
Hartford, CN	65	65	61	52	41	29	27	28	36	48	58	64	47.8	3	6,278
Juneau, AK	55	54	49	42	34	28	25	27	30	38	46	52	40.1	4	9,089
Lansing, MI	65	64	60	51	38	27	24	24	32	46	56	63	46.0	4	6,935
Los Angeles, CA	65	65	65	64	61	58	55	57	58	61	63	64	61.3	0	1,351
Miami, FL	64	65	65	65	63	63	64	64	65	65	65	65	64.4	0	219
Milwaukee, WI	64	63	59	50	36	25	21	22	31	44	53	61	44.0	4	7,665
Minneapolis, MN	64	64	59	50	32	19	14	16	28	44	55	62	42.3	5	8,295
Oak Ridge, TN	65	65	64	59	47	40	40	41	47	57	63	65	54.5	0	3,833
Portland, MA	65	63	59	49	38	26	22	23	31	43	53	61	44.3	4	7,556
Portland, OR	64	64	61	54	45	41	38	42	46	52	57	62	52.2	0	4,635
Prudoe Bay, AK	39	39	28	29	5	9	-16	-24	-16	-8	20	35	11.7	8	19,445
Rapid City, SD	64	65	60	49	35	27	22	24	31	45	54	61	44.8	4	7,373
Salt Lake, UT	65	65	62	51	37	30	27	33	40	50	57	62	48.3	3	6,096
Seattle, WA	63	63	61	54	47	44	41	44	46	52	57	61	52.8	0	4,453
South Bend, IN	65	65	61	53	39	29	26	27	35	48	57	63	47.3	3	6,461
Spokane, WA	65	64	59	49	36	30	25	30	38	47	56	61	46.6	3	6,716
Washington, DC	65	65	64	58	48	38	37	38	45	55	63	65	53.4	0	4,224
Williamsport, PA	65	65	61	53	41	30	29	29	37	49	59	64	48.6	3	5,934

National Climatic Data Center, National Oceanic And Atmospheric Administration, Asheville, NC, 198 months; Fairbanks AK 5 months, Minneapolis MN 5 months; Portland OR none, etc. This data was extrapolated from climatic data recorded by National Oceanic and Atmospheric Administration, [15].

Climatic Conditions.Table 4. In order to determine the amount of water vapor in the air it was necessary to obtain daily temperatures such as: high, low and and dew-point. The dew-point temperature is necessary to compute the amount of water vapor in the air and its driving force. The vapor pressure differential and the permeability of the primary and secondary vapor retarders determines how much water vapor may diffuse into the wall cavity. If the temperature gradients at the vapor retarder remains below dew point the moisture will condense. The condensate will freeze and build up as frost, if the temperature in the cavity remains below 32°F for prolonged periods.

Dew-point temperature is the point where water vapor in the air will begin to condense out as free water, rain or snow. Also it is the temperature where water vapor will condense on any surface whose temperature may be below the dew point of the water vapor in the air. A dew-point hygrometer is an instrument used for measuring the temperature at which vapor being cooled in a silver vessel begins to condense.

The driving force of the moisture in the air was computed in pounds per square feet (Psf) based on the inside and outside dew-point temperatures. The values were computed by the computer. The pressure differential is the difference of inside and outside values. The pressure differential exerted may be positive (outward) during the winter heating season, while it may be negative (inward) during the summer cooling season.

The amount of moisture in the air may be evaluated in two ways. Relative humidity compares the moisture level as percentage of moisture it could hold when fully saturated (100 Pct). Relative humidity can not be used to compare moisture levels of different air temperatures. Therefore, absolute humidity must be used to compare the actual amount of moisture in the air at different temperatures. It is computed as Gr/Lb (grains of water vapor per pound of dry air)   A grain is equivalent to 1/7,000 of a pound.

The Minneapolis outside daily low mean temperatures for the month of January 1988 varied from a low of -18°F (degree Fahrenheit) to a high of 30°F. The mean temperature varied between -10 to 32°F. The dew-point temperature varied from -19 to 30°F. These values were used to compute the absolute humidity in Gr/Lb (grains of water vapor per pound of dry air) . It varied from 2.0 to 24.2 Gr/Lb . The resultant vapor pressure of the inside air as computed varied from 0.9 to 11.7 Psf (pounds per square feet). The vapor pressure differential varied between the inside and outside air varied from 3.9 to 14.7 Psf.

General Design Conditions.Table 5. The outside temperatures were automatically transfered from previous table by the computer. An inside design temperature of 70 F and the relative humidity of 50 percent was entered on the computer key board.

The dimensions of the building and other pertinent data was entered on the keyboard in the shaded cells. Various surface areas and volume of the building was calculated by the computer.

The inside vapor pressure was calculated by the computer as 26.1 Psf. The resultant absolute humidity was computed at 54.0 Gr/Lb.

TABLE 4. CLIMATIC CONDITIONS

Ti =	70	°F, Rhi =	50	Pct, Pvi=	26.1	Psf, Hvi=	70.0	Gr/Sf
Period		Temperatures				Vapor Pressure	Absolute Humidity	Pressure Difference
	High	Low	Mean	Departure	Dew point			
Da	F Tho	F Tlo	F Tmo	F Tdwo	F Tdpo	Psf Pvo	Gr/Lb Hvo	Psf Hvd
Minneapolis, MN, Jan 88								
1	7	-9	-1	-14	-8	1.8	3.6	24.4
2	19	-5	7	-6	1	2.8	5.8	23.3
3	19	-9	11	-1	2	3.0	6.1	23.1
4	3	2	-5	-17	-18	1.0	2.1	25.1
5	-6	-14	-10	-22	-19	0.9	2.0	25.2
6	-1	-18	-10	-22	-18	1.0	2.1	25.1
7	10	-16	-3	-14	-10	1.6	3.2	24.5
8	7	-7	0	-11	-6	1.9	4.0	24.2
9	0	-17	-9	-20	-15	1.2	2.4	24.9
10	20	-12	4	-7	0	2.7	5.5	23.4
11	27	11	19	0	17	6.3	13.1	19.8
12	29	-2	14	3	11	4.7	9.7	21.4
13	9	-9	0	-10	-10	1.6	3.2	24.5
14	22	-6	8	-2	3	3.1	6.5	23.0
15	32	13	23	13	19	7.0	14.4	19.1
16	39	20	30	24	29	11.2	23.1	14.9
17	34	15	25	15	22	8.1	16.6	18.1
18	31	15	23	13	23	8.4	17.4	17.7
19	34	30	32	22	28	10.7	22.1	15.4
20	32	23	28	18	21	7.7	15.9	18.4
21	23	3	13	2	8	4.0	8.3	22.1
22	26	-1	13	1	5	3.5	7.2	22.6
23	12	-6	3	-8	0	2.7	5.5	23.4
24	21	7	14	3	6	3.7	7.5	22.5
25	7	-4	2	-9	-7	1.8	3.8	24.3
26	4	-14	-5	-14	-12	1.4	2.9	24.7
27	19	-12	4	-8	1	2.8	5.8	23.3
28	23	0	12	0	8	4.0	8.3	22.1
29	39	22	31	19	26	9.7	20.1	16.4
30	36	24	30	21	30	11.7	24.2	14.4
31	34	0	17	4	10	4.5	9.2	21.6
Low	-6	-18	-10	-22	-19	0.9	2.0	14.4
Mean	20	1	10	-1	5	4.4	9.1	21.7
High	39	30	32	24	30	11.7	24.2	25.2

TABLE 5. GENERAL DESIGN CONDITIONS

Minneapolis, MN, Jan 88

OUTSIDE DESIGN CONDITIONS:		Low	Mean	High
Temperature, To	F	-18	5	30
Dew point, Tdo	F	-19	10	30
Vapor pressure				
Saturated, Pso	Psf	NA	NA	NA
Vapor, Pvo	Psf	0.9	4.4	11.7
Absolute humidity				
Saturated, Hso	Gr/Lb	NA	NA	NA
Vapor, Hvo	Gr/Lb	2.0	9.1	24.2
Specific volume, Svo	CF/Lb	11.1	11.7	12.4
Weight dry air, Wao	Lb	827	785	742
Weight of vapor, Wvo	Lb	0.2	1.0	2.6
INSIDE DESIGN CONDITION	Unit	Low	Mean	High
Temperature, Ti	F	70	70	70
Relative humidity, Rhi	Pct	50	50	50
Dew point, Tdi	F	50.5	50.5	50.5
Vapor pressure				
Saturated, Psi	Psf	52.2	52.2	52.2
Vapor, Pvi	Psf	26.1	26.1	26.1
Absolute humidity				
Saturated, Hsi	Gr/Lb	110	110	110
Vapor, Hvi	Gr/Lb	54	54	54
Specific volume, Svi	CF/Lb	13.5	13.5	13.5
Weight dry air, Wai	Lb	682	682	682
Weight of vapor, Wvi	Lb	5.3	5.3	5.3
DIFFERENTIALS:	Unit	Low	Mean	High
Vapor pressure, Pvd	Psf	25.2	21.7	14.4
Bouyancy dry air, Wda	Lb	146	104	61
Water vapor, Wvd	Lb	5.1	4.3	2.7
ALTERNATE HUMIDITY CONDITIONS:				
Relative humidity, Rhi	Pct	30	30	30
Vapor pressure, Pvd	Psf	15.7	15.7	15.7
Absolute humidity, Hvd	Gr/Lb	33.1	33.1	33.1
Pressure difference, Pvd	Psf	14.7	11.3	3.9
Relative humidity, Rhi	Pct	40	40	40
Vapor pressure, Pvd	Psf	20.9	20.9	20.9
Absolute humidity, Hvd	Gr/Lb	44.1	44.1	44.1
Pressure difference, Pvd	Psf	19.9	16.5	9.2

DIMENSIONS:

House, 24-0x48-0, 1-story, open crawl, cathedral

Width	Ft	24	Window	SF	115
Length	Ft	48	Door	SF	40
Perimeter	Ft	144	Wall	Sf	997
Floor	SF	1,152	Rise	In	3
Foundation	Ft	4	Run	In	12
Stories	No	1	Wall	SF	1,187
Height	Ft	8	Ceiling	SF	1,187
Window	Pct	10	Volume	CF	9,216

The calculated specific volume of outside dry air varied from 11.1 to 12.4 Cf/Lb (cubic feet per pound). These values were required to calculate the weight of the dry air and the water vapor in the air.

The weight of the dry air in the house was computed as 682 Lb (pounds) while the outside air varied from 742 to 827 Lb, based on equivalent volume of 9,216 CF (Cubic feet) for the house. The buoyancy of the warmer inside air varied from 61 to 146 Lb (pound), which would drive warm-moist air through any holes in the upper portion of the wall or ceiling. Buoyancy (chimney action) of warm air seems to be special problem in high rise buildings.

The water vapor of the inside air weighed 5.3 Lb, while the weight of the outside air varied from 0.2 to 2.6 Lb. The difference of inside and outside water vapor varied from 2.7 to 5.1Lb. Yet, this seemingly small amount of water may create serious condensation problems, if allowed to condense in the wall cavity. If the water vapor froze it would continue to build-up in the wall cavity as ice or frost.

If the inside relative humidity was reduced from 50 to 40 Pct, the vapor pressure differential would drop from a range of 14.4-25.2 Psf to a lower range of 9.2-19.9 Psf, depending on outside conditions. If the inside relative humidity was reduced to 30 Pct the resultant vapor pressure differential range would range between 3.8-14.7 Psf.

Wall Conditions, Tables 6.1 to 6.3. The purpose of these tables was to determine the amount of water vapor that could possibly be retained in the wall cavity. The structural component included a break down of material and its thermal resistance (R) and vapor resistance (D), which were entered in the shaded areas on the keyboard. Resistance to vapor flow was used as it was easier compute than perms. The "Not used" sections was used to accomodate additional material without having to revise the formulae.

Inside and outside conditions were transfered by the computer from previous tables. Again the monthly outside temperatures varied from a low of -18.0 F to a high of 30.0 F, while the inside temperature was 70°F. The outside vapor pressure varied between 0.9 to 11.7 Psf. The inside vapor pressure was 26.1 Psf.

The perm rating of material is based on a vapor pressure of 70.3 Psf. The diffusion of vapor thru the wall was based on vapor pressure differentials at each component.

Paper Back Primary Vapor Retarder Not Functioning, Table 6.1. The asphalt coated paper vapor retarder in the Minnesota house was inadvertently placed on the cold side of the wall instead of the warm side of the cavity.

The total resistance of the section to vapor diffusion is 0.96 (1.042 perm). The temperature gradient at the secondary vapor retarder varied between -14.3 to 31.7°F, depending on outside temperatures and thermal values of each component. The vapor pressure varied from 15.6 to 20.1 Psf. The dew-point temperature varied from 36.5 to 42.1°F, depending on the vapor pressure. Moisture diffusing thru the wall will condense because the temperature gradient is lower than the dew-point temperature.

Vapor diffusion through the wall would vary between 0.213 to 0.373 Gr/ Hr Sf (grain per hour per square foot), based The total amount of water diffusing through the wall section varied from 109 to 191 Lb for the 5 month period, based on 997 Sf of wall area.. It is estimated that 18 to 31 would be

TABLE 6.1. WALL CONDITIONS, PAPER BACKING NOT FUNCTIONING.

Ti = 70 °F, Rhi = 50 Pct, Pvi= 26.1 Psf, Hvi= 70.0 Gr/Sf

Type Material	Resistance		Temperature Gradient, Tg		
	Thermal	Vapor	Vapor Pressure Gradient, Pvg		
	R-value	1/Perm	Vapor Difussion, Vd		
	Hr SF F	Hg/Gr	Temperature Dew Point, Tdp		
Minneapolis, MN, Jan 88	/BTU	Hg SF	Unit	Low	Mean	High
Air film, outside,	0.17	0.00	F	-18	5	30
15 Moh			Psf	0.9	4.4	11.7
			Gr/Hr SF	0.373	0.322	0.213
Siding, hardboard,	0.39	0.00	F	-16.9	5.6	30.5
12" lap			Psf	0.95	4.41	11.7
			Gr/Hr SF	0.373	0.322	0.213
Paper, sheathing,	0.00	0.56	F	-14.3	7.5	31.7
asphalt, wet cup			Psf	0.95	4.41	11.72
			Gr/Hr SF	0.373	0.322	0.213
Secondary vapor	0.63	0.35	F	-14.3	7.5	31.7
retarder, plywood,			Psf	15.6	17.1	20.1
1/2"			Gr/Hr SF	0.373	0.322	0.213
			F	36.5	38.4	42.1
Fiberglass, 3-1/2"	11.00	0.00	F	-10.2	10.5	33.6
			Psf	24.8	25.0	25.4
			Gr/Hr SF	0.373	0.322	0.213
Not used	0.00	0.00	F	62.5	64.5	66.6
			Psf	24.8	25.0	25.4
			Gr/Hr SF	0.373	0.322	0.213
Primary vapor retarder,	0.00	0.00	F	62.5	64.5	66.6
paper backing on fiberglass			Psf	24.8	25.0	25.4
not functioning			Gr/Hr SF	0.373	0.322	0.213
			F	49.1	49.3	49.7
Not used	0.00	0.00	F	62.5	64.5	66.6
			Psf	24.8	25.0	25.4
			Gr/Hr SF	0.373	0.322	0.213
Not used	0.00	0.00	F	62.5	64.5	66.6
			Psf	24.8	25.0	25.4
			Gr/Hr SF	0.373	0.322	0.213
Gypsum board 1/2"	0.45	0.05	F	62.5	64.5	66.6
			Psf	24.8	25.0	25.4
			Gr/Hr SF	0.373	0.322	0.213
			F	49.1	49.3	49.7
Interior film	0.68	0.00	F	65.5	66.7	68.0
			Psf	26.1	26.1	26.1
			Gr/Hr SF	0.373	0.322	0.213
Total Resistance	13.32	0.96	F	70	70	70
Recipocal (1/R or 1/D)	0.075	1.042	Psf	26.1	26.1	26.1
Sum of Italics	1.19		Gr/Hr SF	0.373	0.322	0.213
TOTAL DIFFUSION Months below 32*F	5		Gr/Hr	372	321	213
			Lb/Pd	191	165	109
CONDENSATIOn: Percent of diffusion	16		Gr/Hr	59	51	34
			Lb/Pd	31	26	18

retained in the wall, based on a 16 Pct retention rate extrapolated from the Maine study. Further research is needed in this area.

Paper Back Primary Vapor Retarder Sealed Properly, Table 6.2. In this case it is assumed the asphalt paper backing of the fiberglass had been placed on the warm side instead of the cold side and it was properly air-vapor sealed.

The total resistance of the section to vapor diffusion is 3.46 (0.289 perm). The temperature gradient at the secondary vapor retarder varied between -14.3 to 31.7°F, based on outside and inside temperatures and thermal values of each component. The vapor pressure varied from 14.1 to 15.0 Psf. The dew-point temperature varied from 12.4 to 34.1°F, depending on the vapor pressure. Moisture diffusing thru the wall will condense because the temperature gradient is lower than the dew-point temperature.

Vapor diffusion through the wall would vary between 0.059 to 0.103 Gr/ Hr Sf (grain per hour per square foot), based The total amount of water diffusing through the wall section varied from 30 to 59 lb, while 3 to 5 Lb would be retained in the wall , based on a 9 Pct retention rate extrapolated from the Maine study. Further research is needed in this area.

Polyethylene primary vapor retarder, 6 mil, sealed properly, Table 6.3. A 6 mil polyethylene vapor retarder placed on the warm side of the wall should provide the best performance, if properly sealed. If not properly sealed, it may not function any better than the improperly installed paper backing.

The total resistance of the section to vapor diffusion is 17.63 (0.057 perm). The temperature gradient at the secondary vapor retarder varied between -14.3 to 31.7°F, based on outside and inside temperatures and thermal values of each component. The vapor pressure varied from 11.7 to 12.2 Psf. The dew-point temperature varied from -7.8 to 31.0°F, depending on the vapor pressure. Moisture diffusing thru the wall will condense because the temperature gradient is lower than the dew-point temperature.

Vapor diffusion through the wall would vary between 0.012 to 0.020 Gr/ Hr Sf (grain per hour per square foot). The total amount of water diffusing through the wall section varied from 6 to 10 Lb for the 5 month period. It is estimated that around 1.0 Lb would be retained in the wall , based on a 9 Pct retention rate extrapolated from the Maine study. Further research is needed in this area.

CONCLUSION:

The following conclusions are offered based on the above field studies and computer simulation.

1. Costly condensation-related maintenance problems and structural safety hazards have been inadvertently created in cold climates due to the lack of a vapor retarder standard in building codes. Building codes should be revised to specify a vapor retarder in climatic zones where condensation may become a problem.

2. The reduction in ventilation rates and caulking of air leaks of the exterior skin in an effort to conserve energy has resulted in a higher relative humidity and resultant vapor pressure, particularly human and animal housing. Some type of ventilation and weep holes at the outer skin is necessary to facilitate natural evaporation of condensate, sublimation of frost, and drainage of excess moisture.

TABLE 6.2. WALL CONDITIONS, PAPER BACKING FUNCTIONING.

Ti = 70 °F, Rhi = 50 Pct, Pvi= 26.1 Psf, Hvi= 54.4 Gr/Sf

Type Material	Resistance		Temperature Gradient, Tg			
	Thermal R	Vapor D	Vapor Pressure Gradient, Pvg			
			Vapor Difussion, Vd			
	Hr SF F	Hg/Gr		Temperature Dew Point, Td		
Minneapolis, MN, Jan 88	/BTU	Hg SF	Unit	Low	Mean	High
Air film, outside, 15 Moh	0.17	0.00	F	-18	5	30
			Psf	0.9	4.4	11.7
			Gr/Hr SF	0.104	0.089	0.059
Siding, hardboard, 12" lap	0.39	0.00	F	-16.9	5.6	30.5
			Psf	0.95	4.41	11.7
			Gr/Hr SF	0.104	0.089	0.059
Paper, sheathing, asphalt, wet cup	0.00	0.56	F	-14.3	7.5	31.7
			Psf	0.95	4.41	11.72
			Gr/Hr SF	0.104	0.089	0.059
Secondary vapor retarder, plywood, 1/2"	0.63	0.35	F	-14.3	7.5	31.7
			Psf	5.0	7.9	14.0
			Gr/Hr SF	0.104	0.089	0.059
			F	12.4	21.8	34.1
Fiberglass, 3-1/2"	11.00	0.00	F	-10.2	10.5	33.6
			Psf	7.6	10.1	15.5
			Gr/Hr SF	0.104	0.089	0.059
Not used	0.00	0.00	F	62.5	64.5	66.6
			Psf	7.6	10.1	15.5
			Gr/Hr SF	0.104	0.089	0.059
Primary vapor retarder, paper backing on fiberglass,	0.00	2.50	F	62.5	64.5	66.6
			Psf	7.6	10.1	15.5
			Gr/Hr SF	0.104	0.089	0.059
			F	20.1	26.7	37.0
Not used	0.00	0.00	F	62.5	64.5	66.6
			Psf	25.8	25.8	25.9
			Gr/Hr SF	0.104	0.089	0.059
Not used	0.00	0.00	F	62.5	64.5	66.6
			Psf	25.8	25.8	25.9
			Gr/Hr SF	0.104	0.089	0.059
Gypsum board 1/2"	0.45	0.05	F	62.5	64.5	66.6
			Psf	25.8	25.8	25.9
			Gr/Hr SF	0.104	0.089	0.059
			F	50.1	50.1	50.2
Interior film	0.68	0.00	F	65.5	66.7	68.0
			Psf	26.1	26.1	26.1
			Gr/Hr SF	0.104	0.089	0.059
Total Resistance	13.32	3.46	F	70	70	70
Recipocal (1/R or 1/D)	0.0751	0.28913	Psf	26.1	26.1	26.1
Sum of Italics	1.19		Gr/Hr SF	0.104	0.089	0.059
TOTAL DIFFUSION Months below 32*F	5		Gr/Hr	103	89	59
			Lb/Pd	53	46	30
CONDENSATIOn: Percent of diffusion	10		Gr/Hr	10	9	6
			Lb/Pd	5	5	3

WATER IN EXTERIOR BUILDING WALLS

TABLE 6.3. WALL CONDITIONS, 6 MIL POLYETHYLENE.

Ti = 70 °F, Rhi =	50	Pct, Pvi=	26.1	Psf, Hvi=	54.4	Gr/Sf
Type Material	Resistance		Temperature Gradient, Tg			
	Thermal	Vapor	Vapor Pressure Gradient, Pvg			
	R-value	1/Perm	Vapor Difussion, Vd			
	Hr SF F	Hg/Gr	Temperature Dew Point, Tdp			
Minneapolis, MN, Jan 88	/BTU	Hg SF	Unit	Low	Mean	High
Air film, outside, 15 Moh	*0.17*	0.00	F	-18	5	30
			Psf	0.9	4.4	11.7
			Gr/Hr SF	0.020	0.018	0.012
Siding, hardboard, 12" lap	*0.39*	0.00	F	-16.9	5.6	30.5
			Psf	0.95	4.41	11.7
			Gr/Hr SF	0.020	0.018	0.012
Paper, sheathing, asphalt, wet cup	0.00	*0.56*	F	-14.3	7.5	31.7
			Psf	0.95	4.41	11.72
			Gr/Hr SF	0.020	0.018	0.012
Secondary vapor retarder, plywood, 1/2"	*0.63*	*0.35*	F	-14.3	7.5	31.7
			Psf	1.7	5.1	12.2
			Gr/Hr SF	0.020	0.018	0.012
			F	-7.8	12.7	31.0
Fiberglass, 3-1/2"	11.00	0.00	F	-10.2	10.5	33.6
			Psf	2.2	5.5	12.5
			Gr/Hr SF	0.020	0.018	0.012
Not used	0.00	0.00	F	62.5	64.5	66.6
			Psf	2.2	5.5	12.5
			Gr/Hr SF	0.020	0.018	0.012
Primary vapor retarder, polyethylene, 6 mil	0.00	16.67	F	62.5	64.5	66.6
			Psf	2.2	5.5	12.5
			Gr/Hr SF	0.020	0.018	0.012
			F	-3.9	13.4	31.7
Not used	0.00	0.00	F	62.5	64.5	66.6
			Psf	26.0	26.1	26.1
			Gr/Hr SF	0.020	0.018	0.012
Not used	0.00	0.00	F	62.5	64.5	66.6
			Psf	26.0	26.1	26.1
			Gr/Hr SF	0.020	0.018	0.012
Gypsum board 1/2"	0.45	0.05	F	62.5	64.5	66.6
			Psf	26.0	26.1	26.1
			Gr/Hr SF	0.020	0.018	0.012
			F	50.4	50.4	50.4
Interior film	0.68	0.00	F	65.5	66.7	68.0
			Psf	26.1	26.1	26.1
			Gr/Hr SF	0.020	0.018	0.012
Total Resistance	13.32	17.63	F	70	70	70
Recipocal (1/R or 1/D)	0.075	0.057	Psf	26.1	26.1	26.1
Sum of Italics	*1.19*		Gr/Hr SF	0.020	0.018	0.012
TOTAL DIFFUSION Months below 32*F	5		Gr/Hr	20	17	12
			Lb/Pd	10	9	6
CONDENSATIOn: Percent of diffusion	9		Gr/Hr	2	2	1
			Lb/Pd	1	1	1

3. A relative humidity of 50 percent may be ideal for human health, but it may not be suitable in cold climates for the health of the building. The relative humidity in a building should be as low as practical. Perhaps the best answer in human housing is to provide a heat recovery ventilation system (HRV) that automatically maintains the relative humidity below the dew point of critical surfaces.

4. An increase in relative humidity results in an increase water vapor pressure differentials. Also, the amount of water air can hold increases with the air temperatures.

5. The amount of vapor that diffuses into a wall depends on its vapor pressure and the composite vapor resistance (permeability) of the materials in the wall.

6. It appears that a 6 mil polyethylene is one of the better vapor retarders. Potential air-vapor leaks at seams, electrical outlets, recessed lighting fixtures, electrical wiring, plumbing, fireplaces, chimneys, access openings in the vapor retarder must be adequately vapor sealed with a suitable caulk or gaskets.

7. A wind barrier is not needed to reduce air leaks as the polyethylene primary vapor retarder can perform this function. Also the plywood sheathing may serve as a wind barrier. A wind-rain barrier may be required with horizontal siding that has no plywood sheathing to prevent rain from penetrating into the wall cavity.

8. The retention of condensate (ice) in a wall appears to be dependent on the amount of time the interior of the wall remains below dew point and freezing. Also it is dependent on the vapor permeability and air porosity of the secondary vapor retarder (exterior skin). Further research should be conducted to determine methods of minimizing this phenomena

9. To minimize condensation in wall cavities during both the heating and cooling periods, the dew point of the wall must be kept under better control by a combination of a vapor resistant rigid insulation placed directly under the interior gypsum wallboard and mineral wool insulation in the cavity. A more detail computer simulation would be required to determine the optimum thermal and vapor resistance of each type of material to minimize condensation based on local climatic conditions.

REFERENCES

[1] Rogers, T. S., "Design of Insulated Buildings", The Roberts Printing Company, Toledo, 1951.

[2] ASHRAE Handbook, 1988 Fundamentals, American Society of Heating, Refrigerating and Air-Conditioning Engineers, Atlanta, 1988.

[3] Barre, H. J. and Sammet, L. L., "Farm Structures", John Wiley & Sons Inc., New York, 1958.

[4] Barre, H. J., Sammet, L. L., and Nelson, G. L., "Environmental And Functioning Engineering Of Agricultural Buildings", Van Nostrand Reinhold Company, New York, 1988.

[5] Gratwick, R. T., "Dampness In Buildings", John Wiley & Sons, New York, 1974.

[6] Croome, D. J. and Sherrratt, A. F. G., "Condensation In Buildings", Applied Science Publishers Ltd., London, 1972.

[7] Angell, W. J. etal., "Tri-State Homes: Preliminary assessment of Moisture Problems, Air Tightness and Air Quality", Fifth International Energy Efficient Building Conference, Minneapolis, April 9-11, 1987.

[8] Tsongas, G., "A Field Study Of Moisture Damage In Walls Insulated Without a Vapor Barrier", Final Report For The Oregon Department of Energy, Prepared by Seton, Johnson and Odell, Inc., Consultant Engineers Engineering, with Department of Applied Science, Portland State University, and Applied Social Research, Inc., Portland, Oregon, November 1979.

[9] Carlson, A. R., "Mobile Home Condensation Problems", University of Alaska, Fairbanks, Ak, 1987.

[10] Carlson, A. R., "Wall Design For Cold Climates", University of Alaska, Fairbanks, Ak, 1987.

[11] Carpenter, W. F., "Vapor Barriers For Buildings Having High Moisture Conditions", Bulletin 623, University of Maine, Orono, 1964.

[12] Lindley, J. A. and Lunde, H. A., "Interior Wall Coverings For Moisture Control", Fifth International Energy Efficient Building Conference, Minneapolis, April 9-11, 1987.

[13] Lstiburek, J. W., "Current Air Wall Approach Techniques", Fifth International Energy Efficient Building Conference, Energy Efficient Building Association, Minneapolis, Minn, April 10-11, 1987.

[14] Brooker, D. B. "Mathematical Model of Psychrometric Chart", Paper No. 66-815, 1966 Winter Meeting American Society of Agricultural Engineers, Chicago, December 1966.

[15] National Climatic Data Center, Local Climatological Data, Monthly Summary, National Oceanic And Atmospheric Administration, Asheville NC, 1988.

Author Index

B
Beall, C., 165
Besant, R. W., 92

C
Carlson, A. R., 210
Cole, G. G., 150

F
French, W. R., 64, 79

H
Hamlin, T., 105
Hoigard, K. R., 124

J
Jhinger, H. S., 105

K
Kehrli, D. W., 201
Kudder, R. J., 40, 124

L
Lies, K. M., 40

M
May, J. C., 160
Myers, J. C., 11

R
Rezkallah, K. S., 92
Ruggiero, S. S., 11

S
Schwartz, T. A., 150

T
Tao, Y.-X., 92
Thomas, R. G., Jr., 53
Trechsel, H. R., 138

V
Vassiliades, J. M., 160

W
Williams, B. L., 1
Williams, M. F., 1

Y
Yanoviak, A. C., 182
Yuill, G. K., 105

Subject Index

A

Adhered systems, 64
Adsorption, 92, 105

B

Barrier walls, 1, 11
Brick veneer, 1
Building codes, adequacy of, 182

C

Cavity wall/rain screen wall, 1
Cavity walls, 11, 124
Cladding (See also Exterior insulation and finish systems), 1, 11
Computer simulation
    condensation development modeling, 53
    moisture modeling, 105
    wall sections and moisture, 210
Concrete wall panels, precast, 11
Condensation, 53, 92, 124, 210
    effect on thermal insulation performance, 92
Curtain walls, 11, 40, 182

D

Damage, moisture, 79
Design, door systems, 201
Design, exterior insulation and finish systems, 64
Design, sealant joint, 165
Design, watertight exterior building walls, 11
Detection, moisture, 40, 105, 138, 160
Dew point, 53, 138, 210
Differential testing, pressure, 150
Doors, exterior, 201
Drying, 105

E

Electric fields, to trace leaks, 160
Expansion, 64
Exterior insulation and finish systems, 1, 11, 53
    in situ testing of, 79
    limitations of, 64
    practical use of, 64
    structural integrity of, 79

F

Fan pressurization, 138
Frame/weatherseal interface, door, 201
Frost, effect on thermal insulation, 92

G

Glass fiber insulation, 92

H

Heat transfer, 92
High-rise buildings, 182
Humidity
    absolute, 210
    ratio, 210
    relative, 124, 138

I

Infiltration, 40, 64, 138, 182
Inspection procedures, leak diagnosis, 40
Insulation, 53
    fibrous, moisture transport in, 92

J

Joints
    control, 165
    expansion, 165
    sealant, 11

L

Lawsuits, 182
Leakage, water, 1, 11, 150, 182
 detection, 160
 doors, 201
 diagnosis, 40
 in situ testing, 79
 tests and usage, 150
Litigation, design and
  construction, 182

M

Masonry, unit, 1
Moisture in walls
 control, 124, 165
 damage, 79
 detecting, 40, 138, 160
 expansion coefficients, 165
 flow behavior, 53
 infiltration, 64, 182
 modeling, 92, 105
 profiles, 105
 sources, 138, 150
 transport, 95, 105, 210
  in insulation, 92
Monitoring, psychrometric, 124

O

Organic solvent, 160

P

Permeability, 105, 138, 210
Polyethylene vapor retarder, 210
Polysulfides, 165
Pressure differential testing, 150
Psychrometry, 124

R

Rain screen wall, 1
Roof leaks, 160

S

Sealants
 building, 182
 door system, 201
 joint, 11, 165
  design and performance, 165
Silicones, 165
Solvents, organic, for leak
  detection, 160
Standards, 182
 exterior door systems, 201
 window performance, 150

T

Thermocouples, 124
Tracer gas, 138
Transducers, 124
Transport, moisture, 53, 92,
  105, 210
 mechanisms, 138

U

Urethanes, 165

V

Vapor
 control, 124
 detection, organic, 160
 diffusion, 92
 pressure, 210
 retarders, 53, 210
 sources, 138
 transmission, water, 53
Veneer
 brick, 1
 masonry, 11

W

WALLDRY, computer program, 105
Walls (See also Exterior finish
  and insulation systems)
 barrier, 1, 11
 cavity, 11, 210

cavity/rain screen, 1
curtain, 11, 40, 182
exterior, 53, 124, 150, 182
repair, 11
window, 182
Water (See also Condensation, Moisture in walls, Vapor)
leakage, 1, 11, 79, 150, 182
penetration, doors, 201
repair solutions, 182
sources, 40, 138, 160
test and test usage, 150
vapor transmission, 53
Weather exposure, 150
Weatherseal systems, 201
Wetting, 105
Wind
effects, 150
pressures, design, 79
tunnel studies, 182
Windows, 40
performance standards, 150